Design Thinking

Teun den Dekker

International Edition

Noordhoff Groningen

Cover design: 212 Fahrenheit, Groningen
Cover illustration: Getty Images, Londen
Translation: Rudy Maarsman

Any comments concerning this or other publications should be addressed to Noordhoff Uitgevers bv, Afdeling Hoger onderwijs, Antwoordnummer 13, 9700 VB Groningen or via the contact form at www.mijnnoordhoff.nl.

The information contained in this publication is for general information purposes only. No rights or liability of the author(s), editor or publisher can be derived from this information.

0 / 20

ISBN 978-90-01-75253-8 (pbk)
ISBN 978-0-367-72367-5 (hbk)
ISBN 978-1-003-15453-2 (ebk)
NUR 801

Acknowledgements

Writing a book is not something you do alone. A special word of thanks to my editor and my (certainly not invisible) ghostwriter Annelie Uittenbogaard.

The following people contributed to the content of this book in either major or minor ways:
Davy van Aarssen, Maaike van Assema, Tijn Bakker, Esther Binkhorst, Ada Bolhuis, Dimphy Bordewin, Rachel Camps, Myriam Cloosterman, Vivienne Curvers, Tieg Ebus, Maurits Fornier, Hans Geurts, Brenda Groen, Steven de Groot, Daan de Haan, Froukje van Houten, Remko Killaars, Rudy Maarsman, Petra Prescher, Inge Rijnders, Remko Schuurmans, Anna Snel, Lindsy Szilvasi, Koen van Valderen, Roel Versleijen, Kirsten Werps, Harm Zom, Jouke Zult and Else Zwarteveen.

Spring 2020,
Teun den Dekker

Table of contents

Design thinking is...

'Make things people want instead of making people want things.'

—Geoff Cubitt, CEO Isobar U.S.

When you see the word 'design', do you think of expensive *Off-White* sneakers or *Christian Louboutin* high heels? Or a beautiful yet totally uncomfortable sofa in a luxury hotel? You are not the only one. For many, the word 'design' is synonymous with expensive and exclusive products that excel in their form, technical design and aesthetics. In other words, luxury goods for the elite.

For a long time, 'design' was the exclusive domain of architects, product designers and graphic designers who created new and innovative products – apparently from scratch. These creative professionals follow a seemingly hazy thought and working process, full of sketching, combining, experimenting and prototyping, resulting in often surprisingly innovative products that match the wishes and needs of the customer. This thought and working process is not only useful for making exclusive design products but can be more widely applied in other areas.

This different way of thinking and working, known as design thinking, has been gaining interest. In 2015, the *Harvard Business Review* dedicated a complete edition to design thinking. The caption read: 'It is no longer just for products. Executives are using this approach to devise strategy and manage change.' In an article for this edition, Jon Kolko (2015) writes that design is no longer only for aesthetic purposes but that an increasing number of corporations are integrating design thinking principles in their organizational processes. Besides large commercial corporations, such as PepsiCo, IBM, General Electric, Samsung, Puma and Philips, various municipalities and other social institutions are using design thinking to improve their services and to help solve complex social problems and challenges.

As a result, the idea that people had about designers is changing; they are no longer just creative types hired to make products more aesthetically pleasing and expensive, but instead are now considered as serious partners who help to look at the challenges facing an organization in different ways.

Until now, organizations were used to thinking in solutions when confronted with problems. Preferably, they want a single and clear-cut answer. But the current complex problems they are confronted with are not self-contained, clear and easily defined, they are typically linked to other problems, interdependent, they transcend organizational boundaries and are dynamic. In addition, the needs and wishes of clients and end-users are changing so rapidly that ready-made solutions are not the answer to these ever-changing demands. Creating innovative products and services and using creative problem solving strategies is more and more necessary.

Design thinking can be used for a product, service, technology, strategy, policy, or organization. It does not matter whether this is about setting up a stage for cultural events for the elderly, a potato processing factory needing to expand its market, or an IT company wanting to improve its services. Design thinkers think about complex problems and design challenges in a different way, which can lead to surprising and innovative solutions that meet the needs of the customer.

Wild game

Problems or questions that are both complex and contradictory are especially suitable for design thinking.
As early as 1973, Rittel & Webber wrote about 'tame' problems and 'wild' or 'wicked' problems. Wicked problems can be formulated in different ways and the way they are formulated determines the solution. Wicked problems do not have clear causes and effects because they are closely connected to other problems. In this case, a single solution to solve such a problem is an illusion, according to Rittel & Webber. Perhaps they were clair-voyant because more and more of our social issues and organizational problems today are wicked problems that require solutions in a broad context.

What if we compare wicked problems with wild game? If you get three attempts to catch a tame rabbit, you can just go for it, chances are you will succeed in catching it. However, if you want to catch a wild rabbit in three attempts, it is better to make a well thought out plan first. Otherwise, who is actually the wild game? The rabbit or you running after it?

Who is this book for?

Since the end of the millennium, design thinking has been getting more attention from the corporate world, universities and colleges. Many universities and colleges now include design thinking in their curriculum. Students in many different disciplines can now follow so called minors in design thinking. Service Design, Design 360, Design Without Limits, Design Thinking and Doing and Co Design Studio are just a few examples of minor courses that can help students apply design thinking in their own specific disciplines. There are actually colleges that have based their courses on design thinking. *Design based education* is an example of this.

This book was written for students and employees who would like to apply the principles of design thinking when addressing challenges, problems or complex (social) issues in a different way in their own profession or field.

You shouldn't just read this book. Design thinking is something you must actually do. Reading about it will give you the basic knowledge but doing it will teach you what design thinking can do for you, your field of study or your work. In this book we therefore stimulate action and emphasize learning about design thinking by doing.

How to read and use this book?

You cannot summarize or define design thinking in a catchy one-liner. There are researchers and design thinkers who stress that design thinking is a way of thinking. Others would think of it as a process that describes the way design thinking works. In practice, design thinking is put to use in a project approach. Finally, design thinking is considered to be a collection of useful and directly applicable tools. In this book we are not going to define design thinking in just one way. For us, design thinking is a way of thinking, a way of working, a project approach and a tool box. Every chapter in this book tells the same story but in a different light. Hopefully, after reading this book, you will be able to form your own view of design thinking.

In Chapter 1, *Design thinking is a way of thinking*, we answer questions such as: How do design thinkers tackle problems and challenges? Which fundamental attitudes do they use and which skills must they have?

Chapter 2, *Design thinking is a way of working*, answers questions such as: Which phases and milestones are distinguished in the design process? What is the difference between the 'messy' design thinking cycle and the more structured design process?

Due to the fact that you can only learn design thinking by actually doing it, we will give you a roadmap of a design project in chapter 3, *Design thinking is a project approach*, so that you can actually start practising with design thinking as a way of thinking and working.

Finally, in chapter 4, *Design thinking is a tool box*, we bring together different tools you can use in design thinking.

By the end of this book, you will have thought, worked and practised as a design thinker, and you will be able to apply what you have learned in a (school) project, in your work, or any personal situation, like when a friend asks you for advice. That is also what design thinking is: looking at the world in a new way.

FIGURE 0.1 Visual reading guide

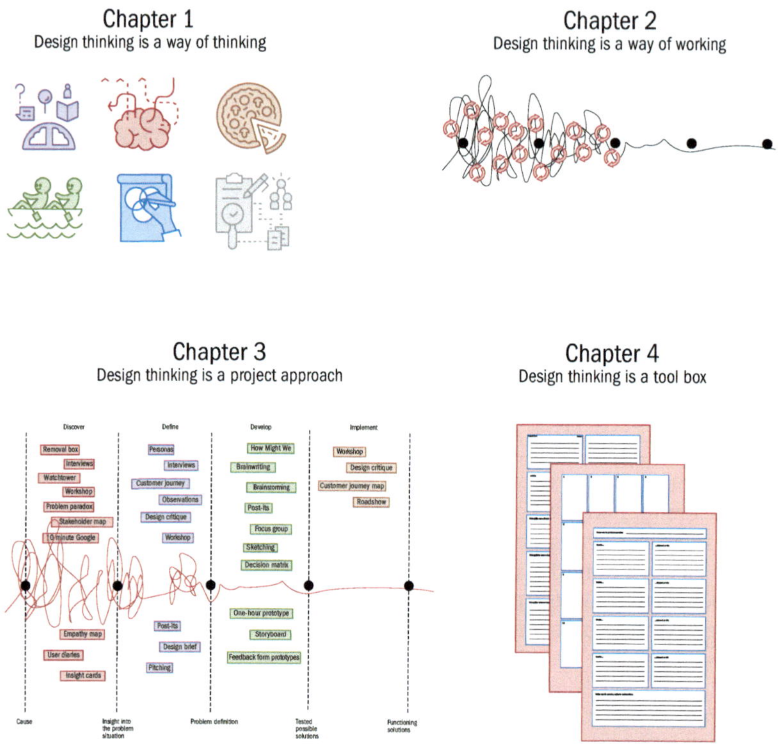

Reading guide

How can you use this book to capture and learn what you think is important?

Here are a few options:

- Chapter by chapter: to discover all aspects of design thinking.
- As a self-study guide, to explore design thinking on your own.
- As a workbook, to practise design thinking: Read chapter 2 and do the exercises in chapter 3.
- As a reference for the design thinker who wants to primarily use the tools in chapter 4.
- As a handbook in your workplace. With this book you can promote design thinking in your company or as an intern.
- Do you only have two hours available? Scan the fundamental attitudes in chapter 1, read through the cycle of design thinking and the design thinking process in paragraphs 2.2 and 2.3 and scan the tools in chapter 4.

DESIGN AS BUSINESS

EXERCIZE

World-wide, there are various companies specialized in applying design thinking to many different issues. These companies call themselves strategic design consultancy firms.

1 Go to the websites of IDEO (ideo.com), Designit (designit.com), Fjord (fjord.com), Livework (liveworkstudio.com).
2 What kinds of projects do these companies do?
3 What distinguishing factors do these companies have that set them apart from one another?
4 How do they sell design thinking?

WHAT'S IN A NAME?

EXERCIZE

Different sources explain and interpret design thinking in different ways. Because of this, concepts have originated which are related or linked to design thinking. Fill in form 0.1 will give you a visual overview of concepts that you may already have come across.

1 Do you know of any other terms or concepts associated with design thinking? Complete the list with your own additions.
2 Google the meaning of the different terms and concepts.
3 Which of these terms and concepts have you come across before in your own field.

FILL IN FORM 0.1 What's in a name

Design based education

Co-creation

____

____

Co-design

Systematic thinking

Interaction design

____

____

Organizational design

UX design

DESIGN THINKING IS ...

____

Human centered design

Innovation

____

Experience design

____

Service design

User experience design

What Design Can Do

In March 2017, the *What Can Design Do Foundation* organized the grand finale of the Refugee Challenge. In cooperation with the United Nations Refugee Organization (UNHCR) they organized a competition for designers from all over the world to think about and address the refugee problem. The idea was to come up with innovative solutions which would improve the lives of refugees in urban areas.

A total of 631 proposals were submitted. Five designers/teams were selected and given 10,000 euros to develop their ideas. Four start-ups emerged from this process. *Agrishelter* makes high quality temporary housing using local materials. *Makers Unite* is a social platform that, together with refugees, develops sustainable products. The *Eat and Meet Bus* is a mobile restaurant that brings together refugees and locals. *The Welcome Card* is an app (among other things) developed with the help of refugees, that strives to make the asylum application process more user friendly.

All these are great initiatives, with the wishes and needs of the customer catered for, or so you would think. Before the results of the competition were announced, a critical article appeared in *de Volkskrant*, a Dutch newspaper. In this article, author Jeroen Jutte, stated he had 'an uneasy feeling' about the competition. He wrote: 'Refugees should not be seen as a problem. This difficult social issue can not be solved by design'. The participating designers were allegedly thinking for the refugees and not with them, and not taking the complexity of these issue into consideration. 'If there is one thing that design cannot do, it is forcing social change', according to Jutte.

Tim Brown, the CEO of the strategic design consultancy firm IDEO has a different take on this. In a TED interview he gave in 2009, he said that design thinking should actually be used to solve complex problems, because by definition design thinkers think up solutions based on the needs and wishes of end-users.

QUESTION 1

Go to www.designthinkinginternational.noordhoff.nl. Read Jeroen Jutte's article (5 minutes reading time) and check out the TED video with Tim Brown (viewing time is 17 minutes).

QUESTION 2

Split into groups (of 2 or 4). Prepare a discussion with one group finding arguments in support of Jeroen Jutte's view and the other group supporting Tim Brown's viewpoint. Then switch roles. Write down your personal opinion about design thinking.

'So I'd like to believe that design thinking actually can make a difference, that it can help create new ideas and new innovations.'

— Tim Brown, CEO and president of IDEO

1 Design thinking is a way of thinking

In this chapter, we will approach design thinking as a way of thinking. Design thinkers have a number of fundamental ways of looking at the world around them and they see different possible solutions for (complex) design challenges. In this chapter we will discuss six fundamental attitudes with which you can apply design thinking in your studies, work or daily life.

We will answer the following questions:
- What are the fundamental attitudes of the design process?
- How do I apply them?
- How do I develop these fundamental attitudes and a general design thinking attitude?

1

Vera Winthagen brings happiness to the neighborhood

Eindhoven is the first city in the world that has recruited a design thinker to combat bureaucracy and compartmentalization using design thinking. In other words, a more creative way of thinking offered by design thinking with the intention to prepare the city for the future.
'I want you to make the city and the municipality happy' said Mary-Ann Schreurs, the city's councilor for innovation, design, sustainability and cultural affairs, to Vera Winthagen on her first day on the job in June 2015. [...]
Design thinking can change the way civil servants do their work, according to the municipality of Eindhoven, who has been hiring designers since 2010. Designers know how to make a product that actually works for the user. Whether it be a coffee machine or a municipal service, the principles used to make a good design are basically the same. For example, designers often work with users as well as with other involved parties and put every idea to the test until everyone is satisfied with the results. Civil servants on the other hand, are used to thinking for the citizens and present solutions only after all details have been ironed out and approved internally.

Go to www.designthinkinginternational.noordhoff.nl to read the entire article and check out which projects Vera Winthagen is doing at the moment.

Source: Het Financieele Dagblad, 19 October 2017 by Ilse Zeemeijer

'Another way of thinking about complex problems is necessary and urgent.'

— Vera Winthagen

1.1 Introduction

If you google 'design thinking', you will see a diverse set of design processes that all describe design thinking in a slightly different way. The processes that are presented look simple and structured. Easy does it, or so you would think.

Appearances deceive. In reality, design thinking tackles problems in a completely different way than the more traditional and linear approaches. There are designers who talk about a certain feeling that is required to arrive at a good design. This so-called 'touch of the designer' is the reason why design is associated with exclusivity and even veiled in mystery. However, the 'feeling' that is needed to complete a successful design process can be analyzed. It appears that there is a clear way of thinking involved in design thinking that can be summarized in six fundamental attitudes, which can be learned.

FIGURE 1.1 The six fundamental attitudes of design thinking

1.2 The fundamental attitudes of design thinking

Design thinking requires a different way of looking at the world. To see the possibilities in the impossible, to make connections that were not thought of before and to work in interdisciplinary teams that challenge everyone to seek solutions for a problem. Design thinking puts the customer first when developing new products and services. In design thinking we believe it is all about channeling imagination and unrelenting experimentation in order to achieve real innovation.

When implementing design thinking, six fundamental attitudes are important:

1 Think flexibly
2 Work integrally
3 Empathize
4 Cooperate
5 Imagine
6 Experiment

Design thinking can conflict with the way we learned stuff at school: linear thinking, working towards a single solution as quickly as possible, thinking about every aspect in a rational manner instead of trying out ideas right away and using your imagination. Therefore, it can take some effort to adopt the fundamental attitudes of design thinking. Our brains simply do not like change and when learning new skills, we first need to unlearn what has become an automatic response. People have trouble changing their behaviors. If you are a regular Instagram user and try not to use it for a week, you will notice how hard it can be not to grab your phone.

Fundamental attitudes

Experience shows that people who start using design thinking learn best by doing. Especially if there is a positive impact on one's word when using the fundamental attitudes, the willingness to learn will be greater. In the following sections, we will give you ample opportunity to practise using the six fundamental attitudes of design thinking, so that you can learn to use these 'automatically'. Table 1.1 summarizes the six fundamental attitudes of design thinking.

TABLE 1.1 The six fundamental attitudes of design thinking

	Fundamental attitude	Applications
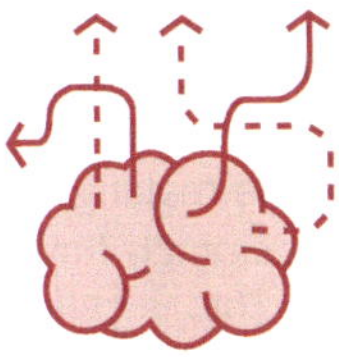	Think flexibly	Balance between: - diverging and converging - analysis and synthesis - zooming in and zooming out - optimism and a critical view
	Work integrally	Become a T-shaped person Look for the innovation sweet spot
	Empathize	Develop empathic ability
	Cooperate	Cooperate in interdisciplinary teams
	Imagine	Learn to visualize Storytelling Creating prototypes for visualization
	Experiment	Learn to experiment Develop an eye for serendipity

1.3 Think flexibly

As a design thinker you will continually want to look at things differently by approaching a problem from all different angles. To look at the big picture, then zoom in on the details, to analyze and then to schematize. This requires your brain to be flexible. People with a flexible brain think up interesting new ideas and solutions to problems that others do not come up with. They have an optimistic and positive attitude without losing sight of critical thinking.

1

1.3.1 Balancing between diverging and converging

Think about a time when you were involved in a project and had to come up with a subject for a group task. A number of subjects were probably proposed pretty quickly but the focus was more on making a choice from a limited number of possibilities. If you want to come up with that one subject that will surprise and be as innovative as possible, you will need to explore more alternatives first. To achieve this, you will need to postpone making choices and focus on creating more alternative choices that seemingly have nothing in common, by brainstorming and making connections. This is called diverging. This entails increasing the number of options to choose from.

Diverging

Converging

Converging is working from several solutions or answers to a single solution or answer. Wanting to achieve this too quickly can lead to a forced choice between options. This could be a bad idea because all options may have very worthwhile elements or characteristics. It can also lead to choosing the best answer at that time which, in the final analysis, is thought to be the right answer. Are you going for the best solution at that moment or are you going for the best solution in the long run?

In the short term, it can seem attractive to come up with a solution by converging without first diverging, but in the long run this will not lead to game changing ideas. Balancing between diverging and converging will require learning to use flexible thinking, thus increasing the possibilities for a solution and knowing when there are enough options on the table from which a solid choice can be made.

FIGURE 1.2 Diverging and converging

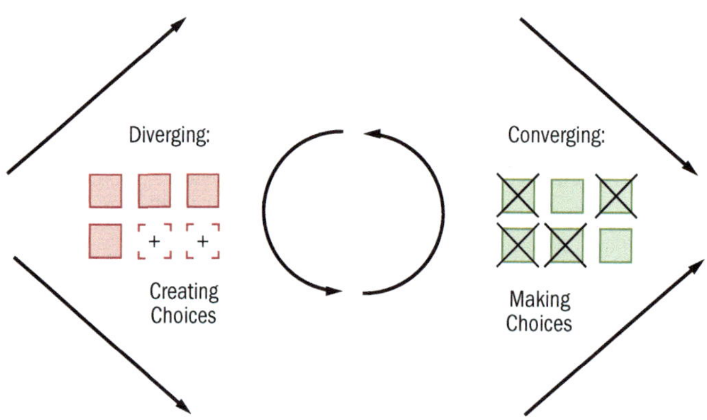

1.3.2 Balancing between analysis and synthesis

It is difficult to understand the world around us without analysis. Analysis comes down to taking something complex apart and then turning it into digestible parts to yield to a greater understanding. This requires you to widen your perspective and gather as much information as possible. By using synthesis, everything that was analyzed can then be interpreted and organized into one story. The raw information from the analysis can be translated into meaningful patterns and new insights. When to use analysis and when to use synthesis is not a question of chronological order. During the entire design process, you will constantly be using both techniques, dissecting the elements of a problem and its solution and then merging them into a larger entity.

Analysis

Synthesis

FIGURE 1.3 Analysis and synthesis

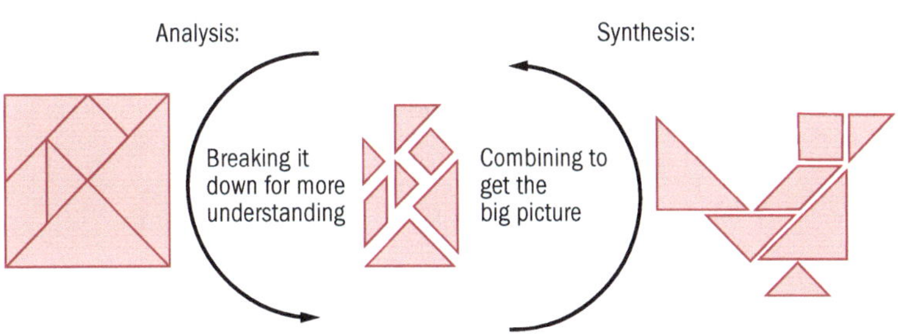

Zooming in
Zooming out

1

1.3.3 Balancing between zooming in and zooming out

Flexible thinking requires a good balance between zooming in and zooming out. When zooming out, you are looking at something from a distance in order to get perspective and overview. That distance is necessary to see relationships and connections so that you can derive logic from these.

Zooming in on the smallest details is just as important. The key to insight into problems and their solutions is often hidden in the details. Professor Rosabeth Moss Kanter (2011) summarizes it as follows: 'Zoom in, and get a close look at select details - perhaps too close to make sense of them. Zoom out, and see the big picture - but perhaps miss some subtleties and nuances.'

A design thinker is constantly switching between the strategic and the operational level, between getting an overview and working out the details. Both the big picture and detailed picture are essential in solving a design problem.

FIGURE 1.4 Zooming in and zooming out

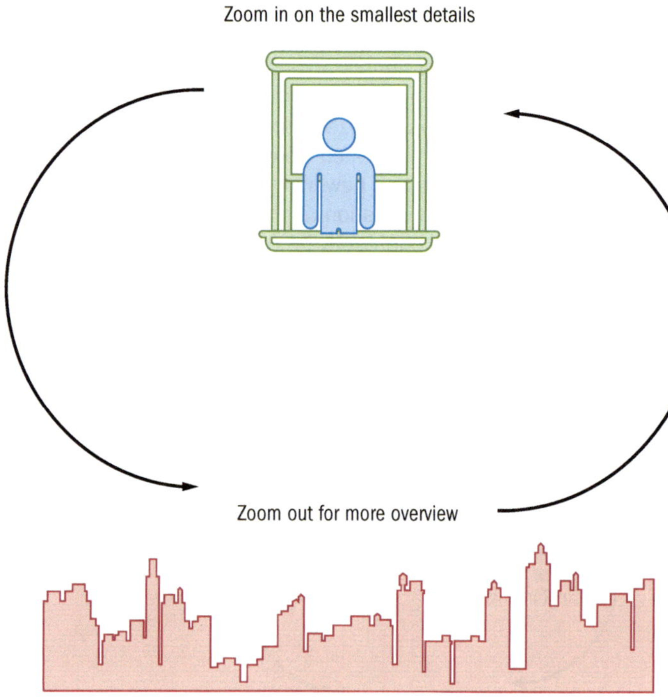

WHAT DO YOU NOTICE?

EXERCIZE

Here we will show you two photographs taken during the mission to execute Osama Bin Laden on May 1st, 2001. The president at the time, Barack Obama, followed the mission in the situation room of the White House.

1 Check out the picture below. Based on this picture, try to imagine what was going on in the situation room.

2 Now look at this next photograph below. Again, imagine how things would have been in the situation room at the time.
3 After seeing the complete picture, did your perception of the situation based on the first picture change?

1.3.4 Balancing between optimism and a critical view

Optimism

Design thinkers are known for their optimism and positive attitude towards problems. Entrepreneur, design thinker and professor Ad van Berlo describes this as follows: 'Designers see a glass as always being half full. That is definitely occupational deformation, I have this myself as well. But in a period of major changes, you have two possibilities: either you hold on to the past or see the changes as a great opportunity to make something new. As a designer I want to move forward.'

The way in which a balance is struck between optimism and maintaining a critical view is closely linked to the design thinker's personality. There are always people who look at new situations and the world around them in a more conservative manner. An optimistic viewpoint means that there will always be a positive solution for a problem or that the situation will develop in a positive manner in the future. A critical view will help to reject assumptions and infuse realism in crazy ideas.
Design thinking requires constantly switching between optimistic and critical viewpoints.

1

'YES, BUT...' OR 'YES, AND...'

EXERCIZE

Part 1: Split into groups of two. One plays the role of the optimist, the other the pessimist.
The optimist starts the conversation and says: 'I want to go to Thailand on vacation!'
The optimist will now have two minutes to try and persuade the pessimist to go with him on vacation (use the timer on your phone!). The pessimist may only react with: 'Yes, but...' phrases.

- Do you think you will go on vacation to Thailand in the end?
- How would you describe the cooperation?
- How do you feel after doing this exercise?
- What ideas did you get?

Part 2: Now repeat the exercise with both of you playing the role of the optimist. Again, in two minutes the optimist from part 1 will try to convince the other to go with him on vacation to Thailand. Optimist number 2 may only react with 'Yes, and....' phrases.

- Do you think you will go on vacation to Thailand in the end?
- How would you describe the cooperation?
- How do you feel after doing this exercise?
- What ideas did you get?

1.4 Work integrally

A problem is rarely self-contained and not easy to define simply and clearly. By the same token, a solution can not be seen in isolation, from a single perspective. Problems and their solutions are generally too complex for that.

Working integrally means cooperating and connecting with others, in order to create coherence. To come to suitable solutions, professionals will need to link their expertise and actions. The complexity of problems and their solutions require creative problem solving strategies that approach the problem and the possible solutions in the broader context in which they occur. This can be successful if one actively looks for the links with related problems, inside and outside the organization.

1

Besides being complex, problems continually change. Design thinking helps to look for solutions that are practical but also take the dynamics of the problem into account. In order to successfully implement solutions, all aspects and cross connections must be considered during the entire design process, especially when determining what the problem is and thinking up solutions.

1.4.1 Becoming a T-shaped person

T-shaped person

Since the eighties, consultancy firm McKinsey & Company talks about *T-shaped persons*. These people have mastered the fundamentals of WORKING INTEGRALLY. T-shaped persons have specialist knowledge in their own specific field but also have general knowledge of the fields their colleagues are specialized in. Combining this knowledge will lead to more effective collaboration when working on solutions.

FIGURE 1.5 Becoming a *T-shaped person*

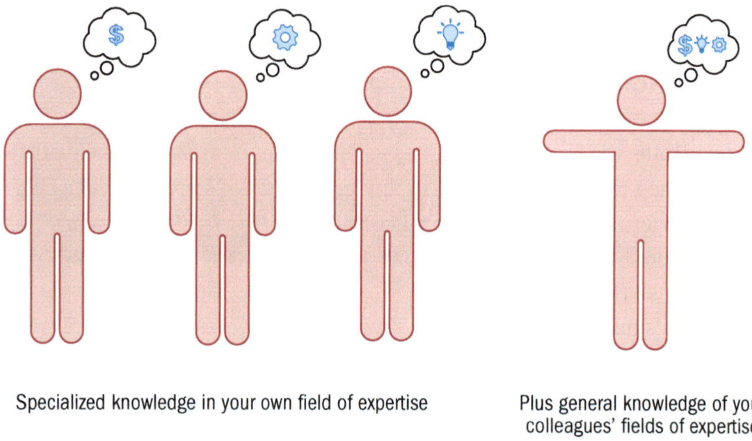

An example will clarify this. Imagine that a group of friends are developing an app. One of them is technically proficient and able to make the app. Another team member is able to translate customer demands into concrete functionalities. Yet another is really good at finding investors and knows how to approach the market. If these friends would only focus on their own areas of expertise, instead of combining their expertise, the app would not be as successful. The messy and iterative design process requires that the friends continually delve into each other's area of expertise so that the technical possibilities are in sync with the functionalities that the customer really needs and with the desires of the investors.

1

1.4.2 Looking for the innovation sweet spot

A design team finds solutions in the innovation sweet spot by considering the following three factors: feasibility, viability and desirability.

Innovation sweet spot

Feasibility is determined by what is available in an organization in terms of technology, budget, staff or partnerships. The important question here is whether the solution will ultimately strengthen the organization.

Feasibility

Viability is all about translating an innovative solution to a sustainable business model, meaning that money can be made from this innovation, now and also in the future.

Viability

Desirability focuses on what the end-user wants and how the design team can ensure that the solution reflects this.

Desirability

FIGURE 1.6 The innovation sweet spot

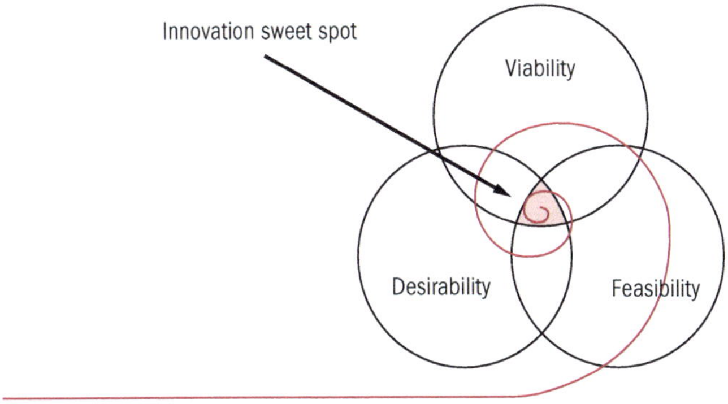

Source: IDEO U (z.d.) (adapted)

1.5 Empathize

To meet the wishes and needs of the end-user as closely as possible, it is this end-user who plays a central role in design thinking. A design solution that does not meet the wishes and desires of the end-user is therefore not a proper solution according to the philosophy of design thinking. The fundamental attitude EMPATHIZE is all about understanding the other person as well as being able to step in the other's shoes to get a feel for their world. This is the definition of empathy.

Empathy

During the entire design process, the design thinker keeps investigating the wishes and needs of the future end-users by empathizing. This can be done via interviews, observation and asking questions via questionnaires. It is only by showing real and structural interest in the customer that the design thinker can gain insight - beyond general assumptions - and can come up with solutions that truly meet the needs of the customers concerned.

● www.limburger.nl, 18-10-2018, by Hennie Jeuken

Students in the world of the patient

Students of Health Sciences at the University of Maastricht immerse in the 'world of the patient' for eight weeks so that they, as future health managers and policy makers, know what happens in daily practice. [...] The students have direct contact with patient associations. In groups of three, they do research for a total of fourteen participating organizations. They learn to communicate with patients and discover which issues and problems these patients are confronted with, for example as a result of policy making. 'This is the young generation that will be at the wheel soon. It is important that the health managers of the future know what is going on in daily practice, because these days the patient is more assertive and is a major stakeholder in health care. We do not want our students in an ivory tower', according to university professor Nynke de Jong. [...]

Developing empathic capacity

There are four phases in developing empathic capacity (Kouprie & Sleeswijk Visser, 2009). 'Entering the user's world - taking user's point of reference - resonate with the user - leaving the user's world' is the mantra for empathy in design thinking:

1 Entering the user's world starts with discovering his world. This can be via a face-to-face meeting, an internet search or reading a research paper. This sparks interest in the other party.
2 Taking user's point of reference means literally leaving your office or classroom and experiencing the other person's world firsthand, get immersed in this world. This can be achieved by walking/cycling/driving to work with the other person, working together, cooking together, etcetera. In this way, you can gain insight into their mindset. Suspending judgment, is absolutely essential in this phase.
3 Resonate with the other person means, as a design thinker, referring to your own memories and experiences. This creates a connection on an emotional level which is needed to understand someone else's feelings.

In addition, this background knowledge is necessary to give meaning to the behaviors, choices and needs of the other person.

4 Detach from the (future) users altogether and distance yourself from them, leaving the user's world. The trick is to transcend the wishes and needs and look at the underlying driving forces of the other person. By taking a helicopter view, patterns can be discovered that contain the core of the problem and offer new solutions.

FIGURE 1.7 The four phases of empathy

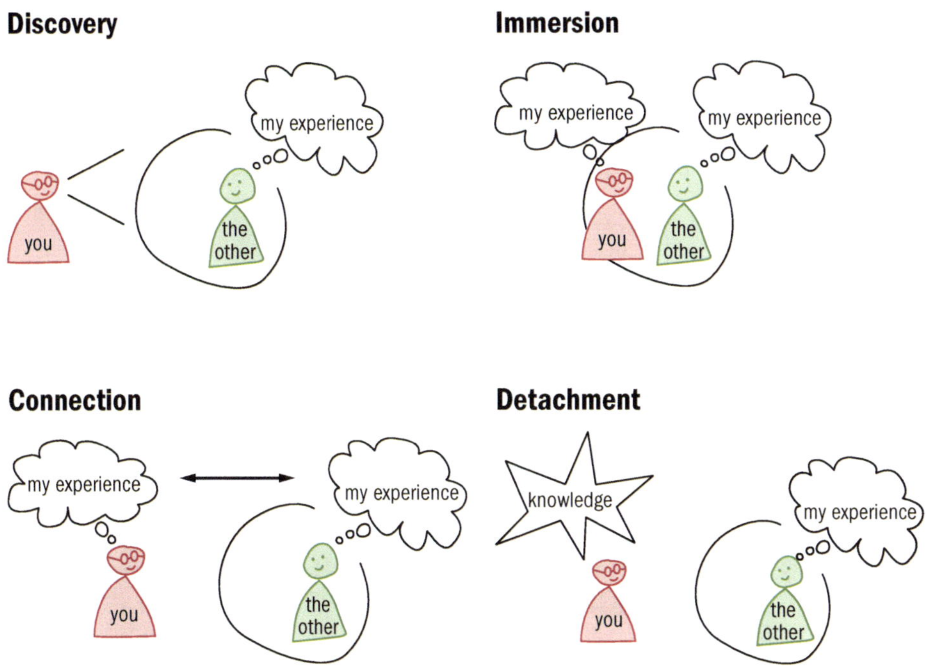

Source: Kouprie et al. (2009) (adapted)

The fourth phase seems contradictory because design thinking puts the customer first but also wants to come up with different kinds of solutions that may not be what the user is actually asking for. However, this is what true innovation is all about: by looking beyond the actual user demands at the underlying driving forces, something can be developed that solves the core of the problem that the customer is experiencing. Henry Ford understood that people wanted to move faster and thought up the production automobile: 'If I had asked people what they wanted, they would have said faster horses.' And Steve Jobs recognized this as well: 'A lot of times, people don't know what they want until you show it to them,' Design thinking uses empathy to develop a service or product that allows users to realize their goals. Do not use empathy to find out what solutions people want but investigate what people feel, think and do and discover why. Translate this into valuable insights and different kinds of solutions.

You will never understand it anyway

Japanese politicians thought that the 'Japanese man' should show more understanding for pregnant women because, of all men in the world, they tend to be the least helpful in the household. With a so-called 'empathy belly' around their waist, the politicians set the example. The video about this went viral. Do you also have no idea what it means to be pregnant? But do you think about problems, solutions, products or services for pregnant women? Or do you want to help your girlfriend or sister during their pregnancy? In that case, you can apply the four phases of empathy:

1 Discovering the world of the pregnant woman starts with a first acquaintance. Visit an informative meeting for pregnant women, check out the products that are for sale, check out the website of a midwife or read scientific research about giving birth.

2 Get immersed in the discomfort of pregnancy. Order an 'empathy belly' (yes, they do really exist) and walk around with it for a day.

3 Connect with what you experience. If you are physically tired, answering the doorbell is already an ordeal. If you want to retrieve a bag of potatoes from the lowest rack in the supermarket, you would fall over. Riding your bike with a full shopping bag is not exactly easy either. By referring to such situations that you have experienced yourself, you will gain more understanding of how a pregnant woman feels and what it means to be pregnant.

4 Finally, by detaching yourself from all this, it will be easier to see how you can support your girlfriend or sister. Relieved to be freed of your empathy belly, you will enjoy the freedom to move freely, stretch and bend again. With this in mind and with a fresh view on the matter, you will suddenly see what your girlfriend or sister needs and how you can develop products and services for pregnant women.

1.6 Cooperate

Design team

Applying design thinking is a team effort. This means that in every design process, a design team is set up in which the team members cooperate to come up with new solutions to (complex) problems. The picture of the solitary designer who comes up with brilliant ideas on his own is herewith discarded.

1.6.1 Get the best out of yourself and others

Cooperation starts with getting the best out of the other team members instead of forcing your own opinions, ideas and solutions on others. It is all about building on the ideas of others. An initial idea can become a truly valid idea via interaction. Without the comments, ad-libs or jokes of a jovial colleague, an idea might never have seen the light.

Cooperation in a design thinking context does not always mean that everything is a group effort. By working separately on a regular basis, everyone can contribute their own specialized subject matter expertise and unique qualities to the design process. These moments of individual thoughts and reflection are needed to effectively contribute to the design process. Besides cooperation within the team itself, a cooperative attitude is needed to involve end-users, subject matter experts, managers, sponsors, creative thinkers, technological whizzkids and other stakeholders. They will all add value to the final solution. In design thinking, cooperation is all about how an individual can have a positive influence on the dynamics within the team, and how the team can involve others in such a way that creative solutions for the design problem can be found.

1.6.2 Successful teams

Most organizations have an organigram depicting their divisions and departments. You can see the hierarchical relationships within the organization and these days there often is a separate innovation department included. Design thinking regards innovation as part of everyone's job and sets up a design team based on relevant expertise, attitudes and competencies. A design team works across the organization. In other words, a separate department or manager is not responsible for solving a design problem but the whole organization is. Design thinking assumes a great deal of autonomy on the part of the design team. Managers who are focused on micro management may find this autonomy challenging. It is important for the design team to keep communicating, not from a hierarchical and accountability point of view but from a feeling of intrinsic motivation to involve others in the design process.

Successful teams

Google's HR department studied the characteristics of the most successful teams within the company. A total of 180 internal team members were interviewed. The generic conclusion was that team dynamics were responsible for a team's success and not the sum total of all the talents of the individual team members. The following five traits were considered the most important for successful teams:

1 **They establish psychological safety**. Psychological safety within a team means that team members dare to take interpersonal risks. They feel free to make contributions without asking themselves whether the

contribution is 'smart, good or relevant' enough. In a 'safe team' an idea which is contributed may be incorrect but is not labelled as 'a failure'. Also, harsh feedback can be given in a 'safe team' without the members feeling they are personally under attack or that the team spirit is compromised.

2 **They require dependability**. In teams it is a requirement that people feel dependent on other team members. This ensures that team members stick to deadlines.
3 **They have a clear structure and clarity**. Team members need to get freedom and trust from their managers, micro management is killing. Within the team, a clear structure and boundaries give each individual team member the space they need.
4 **They give each of their members meaning**. Each team member should be able to add value to the team. This gives team members the energy and enthusiasm for the task at hand.
5 **They have a purpose**. Teams that are convinced that they are contributing to the greater good, show more effort, motivation, enthusiasm and energy in working towards the best solution.

In a successful design team, team members will be different from each other. Therefore, focus on each other's qualifications and know each other's weaknesses. It is each individual's unique specific knowledge and skills that make a strong team. The advantage of having differences between team members is that each member looks at the design problem from their own perspective (based on their knowledge, skills, cultural background and other characteristics).

Teacher Inge Rijnders observes that in practice students prefer to see each other as equals: 'To see each other as equals is easier than having to check out everyone's qualifications and weaknesses'. The ambition of a design team is to create an interdisciplinary team composed of T-shaped persons. That means that there is not only a sum total of the knowledge and skills of the individual team members as in multidisciplinary teams (1+1=2), but that team members also look for synergy in the knowledge and skills of the individual team members (1+1=3).

1.6.3 Work on working together

Working together

You can learn about the fundamental attitude of COOPERATE by adopting a cooperative mindset.

Cooperative mindset

Gorodsky and Rubin (2014) give a number of tips that help to make a team successful:

- **Know each other's strengths**. Be a band of superheroes and contribute your own strengths (specialization) and accept other people's weaknesses.
- **Leverage diversity** by being a pain in the neck. The difficult team dynamics do not outweigh the contribution that your 'being different' makes for the final solution.
- **Get personal** and teach yourself to be a 'complete person' when you are at work. Also share personal matters. This increases engagement with others.
- **Build a relationship** with each other and with the outside world. Do not underestimate the value of a network!
- **Craft your team experience in advance**. Make clear how you want to work as a team: what principles will be adhered to, how members are going to help each other, what the team wants to achieve and what the team is satisfied with.
- **Have fun**! Spend time with the others by going to the pub, playing games, exercising together or doing something else to relieve tension and bring in some humor. The 'wasted' time more than pays for itself.

PAIN IN THE NECK OR PRANKSTER?

EXERCIZE

Think of the times that you have worked together in a team.
1 Which of the tips from Gorodsky and Rubin have you applied before?
2 Which would you like to apply?
3 Could you be both a pain in the neck and a prankster?

Design thinking demands a lot from the cooperation between members of the design team. The team goes through a collaborative process with highs and deep lows. An experienced design team knows that things will ultimately work out. However, a design team with less experience can become quite discouraged, if only because assumptions have to be reviewed time and time again, or have to be revised, or that not all (individual) ideas are accepted by the team. Good ideas may be rejected, prototypes may not work, etcetera. Project members display behavior according to a fixed pattern (see Figure 1.8). In the beginning there is enthusiasm: 'This is what we are going to do!' (1). Then it becomes more difficult than expected with a first dip in the project mood (2). When insights are collected and ideas arise from this, it seems to work anyway (3) and this strengthens the team spirit. At this point in time, things become more concrete and definitive. There is no way back, and uncertainty strikes (4). By maintaining a positive attitude, the team will come out of this crisis. In the end the feeling of 'we are doing well' (5) will prevail.

1

FIGURE 1.8 The project mood graph

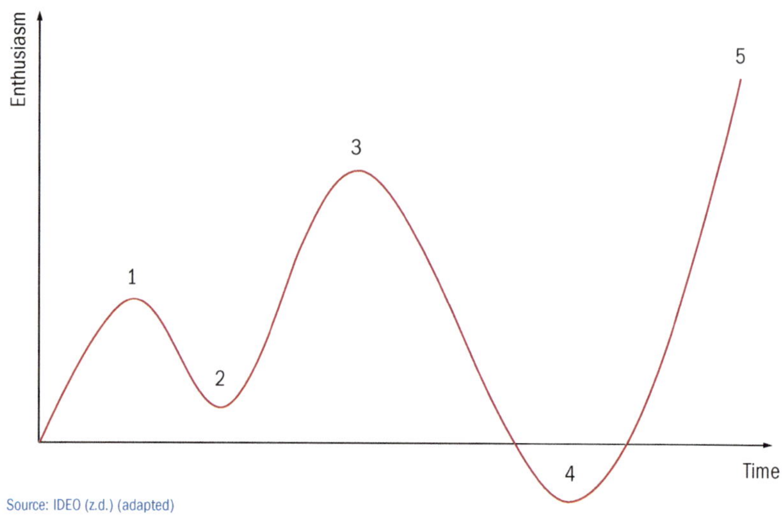

Source: IDEO (z.d.) (adapted)

1.7 Imagine

Whereas kids in first and second grade of primary school may still build and draw stuff to their heart's content, in their subsequent school career the emphasis is placed on finding the right words to understand each other or convey a message. Design thinking preaches: 'Imagination says more than a thousand words.' This fundamental attitude is about: 'Show, don't tell'. Make sure you have sketches, videos, a performance, animations, prototypes and so on to show how an idea will turn out as it continues to be worked out. Let those involved experience the solutions instead of hearing about it. This fundamental attitude is not just about conveying something to the other person, but also helps design thinkers to improve their own thinking about ideas and solutions. This can be achieved by making research data clear in an infographic or by writing down initial ideas that stimulate the creative mind.

1.7.1 Learn to visualize

Whoever uses his imagination will notice that in doing so, a message is more easily transferred, becomes more attractive and is actually better remembered. Visualizing helps to picture information and makes it more expressive. This can be done at any stage of the design process. By taking pen and paper and by using a rough infographic, creating a mind map, drawing some quick doodles or symbols and linking images to an idea, you can use simple visualization techniques that not only help those involved and the team, but also help you with the design process. Kosara (2007) uses the term 'pragmatic visualization' for visualization techniques that generate many ideas quickly, helping to discover alternatives that can initiate a discussion.

Visualizing

Pragmatic visualization

PRACTISING SKETCHING

EXERCIZE

Whoever thinks they can't do anything with pen and paper: *ugly can actually get the job done just fine*. Rediscover the creative child within yourself that drew stuff - without thinking in terms of 'beautiful' or 'ugly'.

1 In five minutes, come up with as many ideas as possible for squeezing one single orange without an orange press.
2 Grab a pen and paper and sketch the ideas you have come up with. Take fifteen seconds each time to make a quick sketch of the idea that:
   a can also serve as a toy;
   b will surprise people on the beach;
   c will surprise people on the train;
   d scares people;
   e makes that you will immediately buy a bunch of oranges;
   f shows what pressing of oranges will be like in the year 2080;
   g ... (come up with a criterion yourself);
   h ... (come up with a criterion yourself).

Was the sketching not so bad or disappointing? Did you gain new insights during the sketching? Did you get completely different ideas?

● www.advocatenblad.nl

Words in drawing

It was no longer possible to make any sense of the case that lawyer Maurits Fornier (32) was involved in a few years ago at Freshfields [one of the oldest international law firms in the world, ed.]. He therefore decided to make an overview drawing and submit it as a procedural document. Fornier is experienced in graphic design. As a high school student he had already made websites for local retailers. 'Look, it is about this and not about all the other subjects', he told the judge based on his illustration. The other party were overwhelmed but did not respond to the illustration and submitted their own plead. After ten minutes the judge interrupted the other party: 'Good, but where are we now in Mr. Fornier's drawing? 'At that moment Fornier knew that he wanted to continue combining imaging and law.

1

Since then, Fornier has continued as a lawyer or legal designer via his own company and is hired by his professional colleagues to provide insight into difficult issues. He recently converted a large legal framework agreement into an A3 poster. He explained provisions that are relevant to employees in the workplace via a single sentence, using pictures and colors. He also made a visual step-by-step plan for an extensive international restructuring. Each step in the plan got a check box as a playful element.
'There was a great deal of coordination work involved for the lawyers, tax specialists and notaries in terms of who did what and what still had to be done. This visual plan was an easy guide for them. Their weekly telephone discussions became much shorter because of this.'

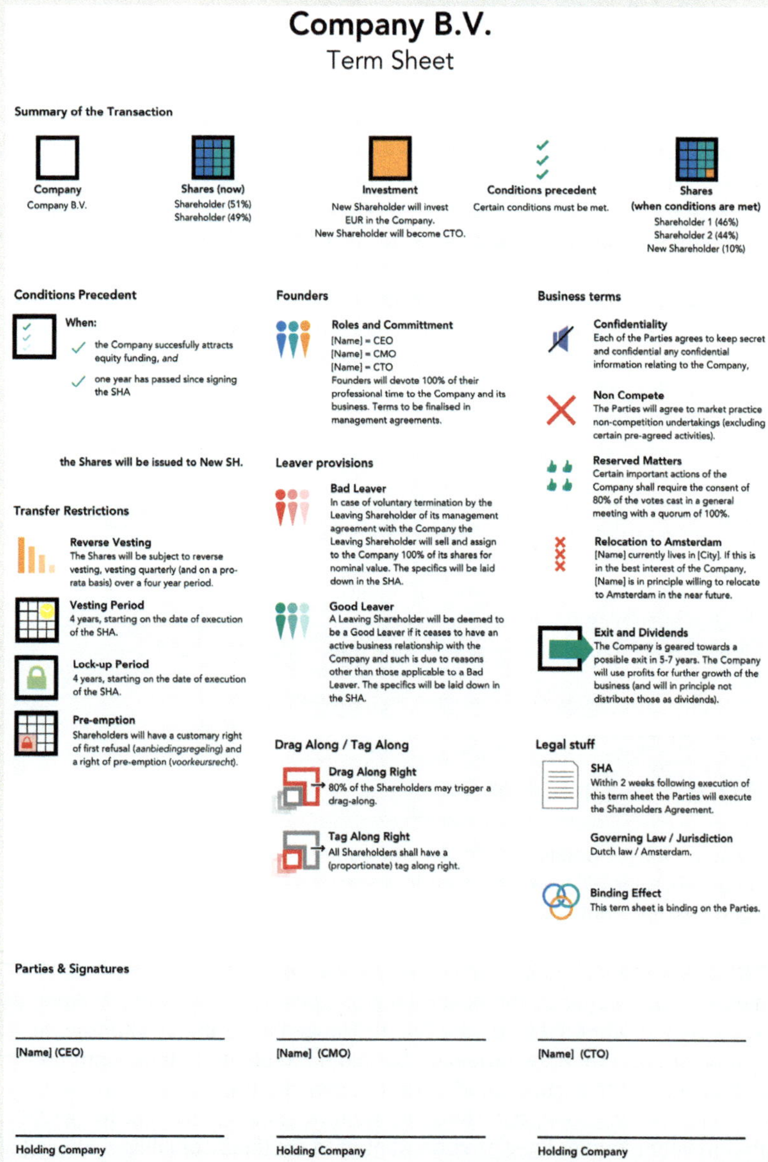

Company B.V.
Term Sheet

Summary of the Transaction

Company
Company B.V.

Shares (now)
Shareholder (51%)
Shareholder (49%)

Investment
New Shareholder will invest EUR in the Company.
New Shareholder will become CTO.

Conditions precedent
Certain conditions must be met.

Shares (when conditions are met)
Shareholder 1 (46%)
Shareholder 2 (44%)
New Shareholder (10%)

Conditions Precedent

When:
- the Company succesfully attracts equity funding, *and*
- one year has passed since signing the SHA

the Shares will be issued to New SH.

Transfer Restrictions

Reverse Vesting
The Shares will be subject to reverse vesting, vesting quarterly (and on a pro-rata basis) over a four year period.

Vesting Period
4 years, starting on the date of execution of the SHA.

Lock-up Period
4 years, starting on the date of execution of the SHA.

Pre-emption
Shareholders will have a customary right of first refusal (*aanbiedingsregeling*) and a right of pre-emption (*voorkeursrecht*).

Founders

Roles and Committment
[Name] = CEO
[Name] = CMO
[Name] = CTO
Founders will devote 100% of their professional time to the Company and its business. Terms to be finalised in management agreements.

Leaver provisions

Bad Leaver
In case of voluntary termination by the Leaving Shareholder of its management agreement with the Company the Leaving Shareholder will sell and assign to the Company 100% of its shares for nominal value. The specifics will be laid down in the SHA.

Good Leaver
A Leaving Shareholder will be deemed to be a Good Leaver if it ceases to have an active business relationship with the Company and such is due to reasons other than those applicable to a Bad Leaver. The specifics will be laid down in the SHA.

Drag Along / Tag Along

Drag Along Right
80% of the Shareholders may trigger a drag-along.

Tag Along Right
All Shareholders shall have a (proportionate) tag along right.

Business terms

Confidentiality
Each of the Parties agrees to keep secret and confidential any confidential information relating to the Company.

Non Compete
The Parties will agree to market practice non-competition undertakings (excluding certain pre-agreed activities).

Reserved Matters
Certain important actions of the Company shall require the consent of 80% of the votes cast in a general meeting with a quorum of 100%.

Relocation to Amsterdam
[Name] currently lives in [City]. If this is in the best interest of the Company, [Name] is in principle willing to relocate to Amsterdam in the near future.

Exit and Dividends
The Company is geared towards a possible exit in 5-7 years. The Company will use profits for further growth of the business (and will in principle not distribute those as dividends).

Legal stuff

SHA
Within 2 weeks following execution of this term sheet the Parties will execute the Shareholders Agreement.

Governing Law / Jurisdiction
Dutch law / Amsterdam.

Binding Effect
This term sheet is binding on the Parties.

Parties & Signatures

[Name] (CEO)	[Name] (CMO)	[Name] (CTO)
Holding Company	Holding Company	Holding Company

A legal contract by Maurits Fornier

1.7.2 Storytelling

Storytelling

If you still want to use words, package the message in an attractive, personal or metaphorical story that appeals to the imagination. Storytelling is the use of stories to depict something. So make sure you have a good story, such as in this radio advertisement: 'You are on the train on your way to work. You hear someone calling ... No, there is no one on the phone, it is a Chinese conductor asking: Do you want a warm towel for your face? Because you are not on the Intercity train to work, you are sitting in the Trans-Siberian Express, to Beijing. You open the window...'

Figure 1.10 shows the structure of a good story. The story starts with a sketch of the situation and the introduction of the main characters. The story is triggered by a so-called trigger moment: something dramatic or unexpected happens and the tension builds up to a climax, (preferably based on a number of sub-climaxes), after which the tension is reduced again. In storytelling, it is important to carefully consider when and which information will be given. We are used to asking ourselves ('which information do I want to send?'), while the question should be: 'what information does the listener need and at what time, to fully understand the message, and more importantly, to embrace it?'

FIGURE 1.10 The structure of a good story

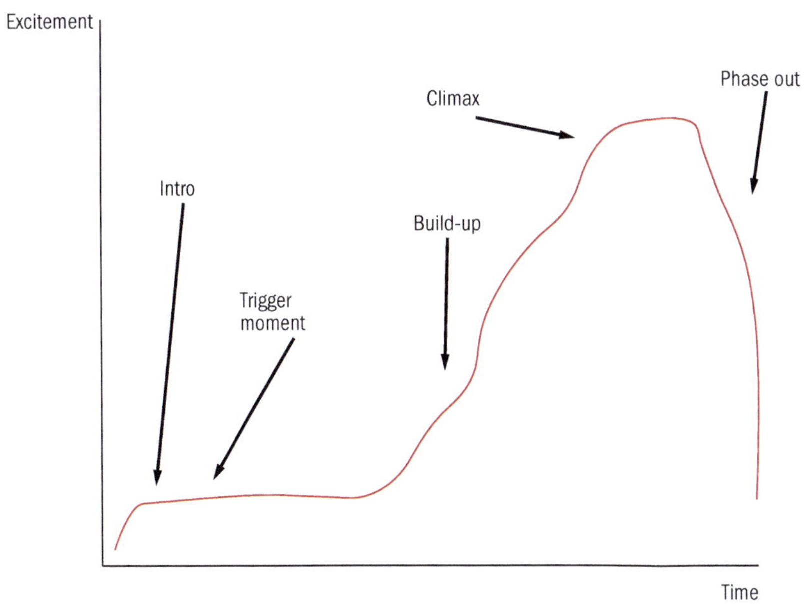

Source: Nijs & Peters (2002)

Liebrecht (2018) elaborates on this and summarizes the theory of Sanders (2017) by identifying the following elements of a good story:

- The story consists of actions and events.
- One or more people are present who are working towards a plot.

- The story is written in the first person; this has more impact than writing in the third person.
- It contains direct quotes; citing is more effective than paraphrasing.
- Time changes take place in the story; switch between the present and the past.

GOOD STORY

EXERCIZE

1 Read the Liebrecht blog and watch the video on www.designthinking international.noordhoff.nl.
2 What steps for building a good story do you recognize in the video?
3 Which elements of a good 'Sanders story' do you recognize?

1.7.3 Creating prototypes for visualization

A prototype represents the elaboration of a concept into something concrete. This can be a visualization, but also a 3D model, role play or (a partly) functioning piece of software. Prototypes are not only intended to give a concrete elaboration of a concept created by users and others involved, but can be tested as a possible solution for the problem(s) or issue(s) that are experienced. Prototypes can also inspire the team to come up with other ideas by physically building something together. Tim Brown (2008) calls this 'build to think'. In making prototypes, design thinking asks you once again to go back to your childhood in which you touched everything, built endless castles with sand, twigs and water or designed your own world with Lego. It didn't bother you if you had to start over, again and again, if for example, the sea washed away your structure; you stayed with it and used your imagination to make something you could see, hear, smell and touch. Making prototypes right from the first phase of the design process, ensures that the design team can fail quickly and cheaply. Making a prototype forces the team to split a solution into smaller parts to be tested. By analyzing and synthesizing each part separately and by zooming in and out, 'mistakes' can already be discovered and corrected at an early stage. It is also possible that via this process, entirely new concepts arise, which in turn, can be worked out in a prototype.

Think carefully in advance about what you want to achieve with a prototype. Also remember what you need: sometimes a partial prototype will already give you sufficient information about an idea or a solution.

Guido Stompff (2018) uses the following guidelines for this:
- Is the prototype primarily intended to convince others? Make the prototype more beautiful than in real life (often used by architects in the form of a scale model).
- Do you want future users to test solutions with the prototype? Make the prototype understandable and functional (software that is tested by frequent users).
- Does the design team have to make a decision about the details of the concept? Ensure that the prototype shows the essence, i.e. the concept(s) behind the solution.
- Does the team want to explore solutions? Make simple and quick sketches to depict the different concepts.

Read more about prototyping and how you can make your prototypes as realistic as possible on page 69.

Chip in your arm

It is good to think in advance which level of precision would be best for the goal that you want to achieve with a prototype. In a design project an idea was developed to enable students to pay for all kinds of services on the college campus using a chip in their arm. The idea was worked out by building an extensive prototype using a mannequin and a technical representation of the chip to make it possible to role play all kinds of situations. The aim was to determine which practical barriers a user would come up against if he wanted to pay with a chip in his arm. After the complicated exercise with the mannequin and the technical representations of the chip, the design team discovered that sticking a simple sticker on someone's arm, to simulate paying for services in real life, was more efficient and effective in testing the idea.

1.8 Experiment

Experimenting is learning by doing. It starts at the beginning of a design process and ends after successfully implementing the solution for the design problem. This fundamental attitude is based on making mistakes, falling, getting up again and giving 'crazy' ideas a chance. This principle can clash with an (organizational) culture in which linearity, certainty and avoiding errors, are highly regarded.

'Failure is not the opposite of success, it is an integral part of success.'

— Arianna Huffington, co-founder and editor-in-chief of The Huffington Post

1

The myth of the never failing genius

The audience applauds for a great piece of music from a brilliant musician or for a Nobel prize-winning scientist. People often believe these are exceptional talents or smart people who can turn everything they touch into gold. Simonton's research (1999) shows that the opposite is true: these talents fail a great deal. The difference with 'normal' people is that they are not afraid to try something out and do not stop when it appears that something is not working. Their brilliant musical pieces or scientific finds do not arise because they 'achieve instant success' but simply because they experiment a lot. If you want more success, prepare yourself for throwing away more failed experiments.

Experimenting offers many benefits:

- It clarifies your own thinking if your assumptions prove to be incorrect based on the results of the experiment.
- It does not have to cost a lot, but not experimenting can cost an organization dearly. By not taking the time to experiment and working out an idea extensively in advance, solutions can be brought to the market where they are not accepted by the customer. Early recognition of errors in our way of thinking or working ensures that the design team can adapt and start again. Failing fast is good!
- Experimentation reduces uncertainty and the sense of risk. Successful experiments confirm that the design process is correct but 'failed' experiments may deliver an even greater source of information. Even if he fails, experimenting always provides the optimistic design thinker with something.

'Don't worry about failure: you only have to be right once.'

— Drew Houston, founder and CEO Dropbox

Return on learning

At ABN-AMRO bank, the term *return on learning* was introduced for experiments which do not succeed, but do provide a learning experience. These learning experiences are widely shared because of the increased learning effect in the organization. This is done via various design teams, internal communication and so-called *return on learning events*. 'Those are actually parties where we celebrate failed experiments, in addition to successful ones, of course' said Tessa Mulder, design thinker at the bank.

Design thinking is impossible without an experimental attitude. An experimental attitude is a playful, curious attitude, but also a critical attitude where you are not easily satisfied and you do not accept the first option that comes up. 'If I can come up with one idea, then there are probably more' and 'you learn the most from mistakes' are typical design thinking attitudes.

Serendipity

In their book *Creative Confidence*, the Kelley brothers write that for experimenting it is especially important to cultivate serendipity. Serendipity means that you find something valuable without looking for it. Pek van Andel calls this phenomenon 'unsolicited finds'. The French chemist Louis Pasteur said: *'Le hasard ne favorise que les esprits préparés'* (chance favours the trained mind). The Kelley brothers summarized this perfectly, coincidence prefers people who do a lot of experiments while keeping an eye out for the unexpected. So an unsolicited find is different from pure coincidence: those who don't pay attention, don't see the opportunities. Serendipity is a choice!

A Nobel Prize by accident?

History has many examples of serendipity. Charles Goodyear discovered how vulcanization worked because he accidentally spilled a combination of rubber and sulfur on an oven and Post-it notes came from a non-sticky glue that turned out to be very suitable for temporary sticking applications. In 1945 Alexander Fleming received the Nobel Prize for Medicine for his 'discovery' of penicillin. Fleming had conducted research into staphylococci (bacteria) and cultivated them for microscopic examination. After a short holiday break, Fleming discovered that one of the petri dishes with bacteria had remained uncovered and that on this particular dish a blue-green mold was formed. The mold that emerged turned out to have secreted a bacteria killing substance: the discovery of penicillin was a fact.

Peka Kroef

1

If you buy a potato product in a Dutch supermarket, chances are they probably come from one of the Peka Kroef factories. Peka Kroef is based in Odiliapeel and is one of the largest potato-processing producers in The Netherlands. The company does not only supply to supermarkets, but through wholesale, also to other companies and restaurants. Peka Kroef saw growth opportunities for this market and their market analyses showed that sales to professional chefs were consistently lacking. To check out all options and not immediately get into a solution mode, a design process was started with the main question: How can Peka Kroef better cater to the actual wishes and needs of the user? This way of troubleshooting needed some getting used to. Harm Zom, Sales Manager Foodservice & Industry at Peka Kroef: 'Design thinking differs a lot from the way we used to work and was initially received with skepticism. In the course of the project we started to really understand design thinking.' The interdisciplinary design team (sales, marketing and product development) that was set up, worked together intensively and had one common goal: understand the professional cook on a deeper level and come up with integral solutions to his problems.

'By involving different people whom we initially would not have thought of, we got a lot more support and valuable information from different angles'. Initially, the design team thought they knew enough about customers and end-users to be able to stimulate sales to professional chefs. But the team was surprised by new insights after having conversations with focus groups and observing professional chefs. The chefs were continuously followed during cooking in order to really be able to relate to them: what the chefs do, think and feel like when they are on the job. For example, if fresh potatoes were available, these were almost always preferred. Instead of using ready-made mashed potatoes, the chefs preferred to puree prepackaged potato slices themselves. Cooking potatoes was something they could do blindly but what about using pre-processed unpeeled potatoes? And why would they use them? The wholesaler delivered new products several times a week and they did not have space for a large stock of potatoes anyway. Contrary to what Peka Kroef had assumed for years, shelf-life of the products was not so important as expected, but packaging was. Harm Zom: 'We thought we knew it all beforehand but the outcomes of the focus group and the observations were extremely surprising. At the end of the day, some persistent assumptions we had about how the end-user thought about shelf-life and packaging were dropped.' The assumptions that proved false had been the basis for decisions that were made in the past and the reason why this market segment had lagged behind.

QUESTION 1

Which of the six fundamental design attitudes do you recognize in this case?

QUESTION 2

Do you miss some fundamental attitudes? How could the design team have used them to come to the end result?

'It was smart using a certain way of asking questions and using the design thinking process to achieve the end results we got.'

— Harm Zom, Sales Manager Foodservice & Industry Peka Kroef

2
Design thinking is a way of working

In this chapter we will look at design thinking as a way of working. We summarize this way of working in a design process consisting of phases, steps and milestones.
Within the more structured design process there is also a 'messy' iterative cycle, in which the six fundamental attitudes are employed.

Read this chapter if you want answers to the following questions:
- What is the difference between the structured design process and the 'messy' iterative cycle of design thinking?
- How does the cycle of design thinking work?
- Which phases and milestones can be distinguished within a design process?
- How do I learn to deal with the uncertain and messy start of a design process?
- What could I come across when I have started a design process?
- What challenges does an organization face if they want to implement design thinking organization-wide?

Julius Caesar and emperor Augustus could not do without 'design thinking'

2

'Design thinking is fashionable,' notes Shanks. 'It is big business.' However, the term is often misinterpreted and misused. Shanks: 'The concept of design thinking is confusing. Because it is not design and actually not a way of thinking either. If you adapt a few parts of your company after having had a design thinking workshop in some creative space, you will not end up with a completely different company. Nor does design thinking offer a solution for all problems. I mistrust advisors who tell me how it should all be done and what the future looks like. Businesses who believe that will fail. [...]' What is design thinking then? Shanks sees design thinking as a form of project management. 'It is a way of acting, a way of working. It is about the way people cooperate in a project. That is why this way of working is so attractive for companies. With design thinking you can organize people in teams so that you can come up with creative ideas and get started quickly. So it is not about solutions, but is a way to move forward. It is about delivering products, services or experiences to people who need it. [...]'

Read the entire article? Go to www.designthinkinginternational.noordhoff.nl and read, among other things how Michael Shanks applies design thinking in archeology.

Source: Het Financieele Dagblad, 29 April 2017, by Ilse Zeemeijer

'It is a way of acting, a way of working.'

— Michael Shanks

2.1 Introduction

The term design 'thinking' is probably created by researchers who were initially focused on how creative professionals *think*. But all six fundamental attitudes that reflect this design thinking mindset all stimulate *action*. The term *design (by) doing* would therefore have been equally applicable.

Design thinking as a way of working is based on a design process. The design process consists of phases, steps and milestones in which we work via a structured approach towards a functioning solution that can actually be implemented.

In this book we make a distinction between a design process of phases, steps and milestones and an underlying, iterative cycle. Other authors, including Gloppen (2009), Norman (2013) and Herfurth (2016), do this as well. The iterative cycle is an integral part of the design process and shows you *how* to use it. It shows how the fundamental attitudes of design thinking interact. It is a continuous learning process that drives the design process from the initial outset to an actual functioning solution for a design problem. It is precisely this underlying cycle, which Stompff (2018) calls 'the cycle of design thinking', that makes design thinking so interesting. The cycle of design thinking shows you how by iterating – continuously running through the cycle until a milestone is reached – you can go on to the next phase and steps of the design process. The way of working of design thinking therefore consists of a design process and *within it*, a cycle of design thinking. Figure 2.1 shows how the design process and the cycle of design thinking are connected and how they interact.

FIGURE 2.1 The design process and the cycle of design thinking

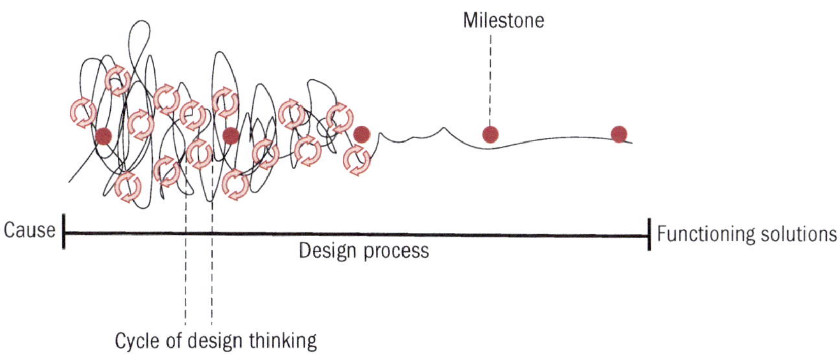

In section 2.2 and 2.3 we will go into the cycle of design thinking and the design process in more detail.

What does gaming have to do with design thinking?

A game consists of different levels (like the phases in the design process); each having a clear goal (milestone). In every level different tasks must be performed before going to the next level. For that, you have to learn new skills. This works via trial and error, by trying, learning and trying again (the cycle of design thinking). By using the acquired skills, you are able to complete more and more levels and eventually you can complete the entire video game.

In the popular game Fortnite you continuously experiment. In the *Save the World* mode you have to complete different missions. Per mission you learn how to perform the different tasks to complete the mission. Imagine that you have to do something that you have never done before, like driving a golf cart. Based on knowledge you already have, you have try driving the golf cart. For example, you assume that 'it will drive by pressing X'. You perform this action, but nothing happens. You adapt your assumption in a split second. You try pressing Y. But again nothing happens. You continue to investigate based on your assumptions until you succeed. Maybe you will try driving while standing, but it turns out that the golf cart will only move when you're sitting behind the steering wheel. Chances are that, within the ten seconds you needed to get the golf cart up and running, you made, tried and rejected twenty assumptions until you discovered how it worked. If you have to build a house in a higher level and you use your golf cart as a means of transport for the building materials, all you have to do is concentrate on learning how to build a house. The cycle of design thinking works similarly: you apply the knowledge and skills you have acquired in the subsequent phases of the design process. The cycle of design thinking can be compared to the continuous learning process within the levels; making assumptions, trying, rejecting and trying again. You can compare the design process to going through different gaming levels and their respective milestones.

PLAY A GAME

Search Google for a website to play Sokoban online. Play at least five levels (or more if you like).

EXERCIZE

1 What did you do when you played the first level?
2 What did you learn and apply in later levels?

2.2 The cycle of design thinking

2

The cycle of design thinking shows you how you can achieve milestones by iteration that will lead to the next phase in the design process.

Iteration

2.2.1 Iterative versus incremental working

Cycle of design thinking

Going through the cycle of design thinking is based on experimentation, whereby ideas are formed and become more specific in every cycle as they are tested before continuing. Sometimes going through the cycle of design thinking takes no more than a split second in your head, other times it takes longer. By making ideas concrete and testing them in every step or phase, weak ideas will be dropped and the best remain. The opposite of this iterative way of working, an incremental approach, is often used in practice. When working incrementally, you work towards a complete solution in a step by step manner. Jeff Patton (2008) uses the example of making a painting. Figure 2.2 shows the difference between iterative and incremental working.

FIGURE 2.2 Iterative versus incremental

2.2.2 Back in time

Iterating with the cycle of design thinking is the most characteristic feature of the design process. This cyclical way of working is anything but new. In *Method for Scientific Research* by Francis Bacon from 1620 (!) a similar cycle had already been described, which was later developed by Popper (1959). The scientific method consists of three steps: hypothesis, testing and evaluation. First it is determined what to investigate. The insights that have been collected are translated into a hypothesis: which is an assumption about the expected answer that has not yet been proven. In the

Scientific method

Hypothesis

second step of Bacon's method the hypothesis is tested, after which in the third step, the results are evaluated by comparing them with the original hypothesis. If the assumption is incorrect, the hypothesis is rejected and a new hypothesis must be formulated and the cycle starts all over again. This process repeats itself until the hypothesis is no longer rejected. The hypothesis is then apparently the best explanation for the facts at that time.

Iterative cycle of human-centered design

Bacon's method is still being used in research. Design thinking uses the cyclic principle in an adapted form. Don Norman (2013) calls this application in design thinking the *Iterative cycle of human-centered design*. He distinguishes four steps: observation, idea generation, prototyping and testing. We learn from him that design thinking is not only about testing and analyzing the hypothesis, but that new ideas will arise continually (idea generation) and that these ideas are worked out and made concrete at an early stage (prototyping). Guido Stompff calls this iterative cycle, the cycle of design thinking, in his book *Design Thinking: Radical Change in Small Steps* (2018). According to him, the cycle of design thinking distinguishes five steps: framing, analysis, idea development, realization and reflection.

Figure 2.3 shows you how each cycle starts with an assumption (framing) that is productive and an incentive to action. You analyze the chosen frame, after which you develop ideas. You work out the ideas in a sketch or prototype (realization), after which your team and the end-users reflect on the chosen frame and make another assumption where necessary (reframing). The cycle of design thinking repeats itself until you have reached the milestone and are satisfied with result. The further you are in the design process, the more concrete the so-called frames become.

FIGURE 2.3 The cycle of design thinking

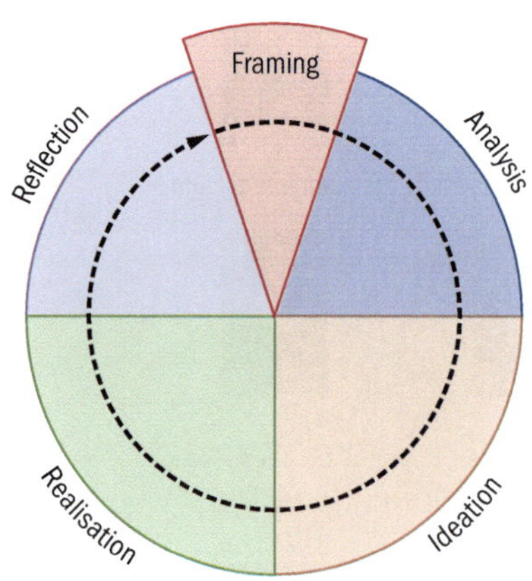

Source: Stompff (2018)

2.2.3 Framing and reframing

What makes the cycle of design thinking so relevant is the first step of the cycle: framing. When Stompff defines frames, he quotes Karl Weick (1995): '...perceptual frameworks that categorize what we see and what we know and that guide our conscious thinking.' **Framing**

Framing is about putting on a certain pair of glasses and taking on a belief or assumption. Reframing is based on following the cycle through findings and adjusting the frame until it is no longer based on assumptions, but on facts. Because there is not one single truth and therefore not one single vision about the problem or a single option for a solution, you keep framing and reframing throughout the entire design process. As a team you ask questions about what the possible problem is, what the possible solutions are, but also about what customers think and what can be achieved in the market. The answers to these questions can only be found through continuously and actively starting a dialogue as a team and with others. By continuously framing and reframing together, an assumption is viewed and tested from all angles. This is how you prevent continuing with a frame that does not contribute to the design process or which customers later see very differently. **Reframing**

Charli or Jillz?

It took Heineken three years to launch the new 'women's drink' Charli. At the end of the process, the drink was tested in 15 cafés. It soon became apparent that the Charli name did not appeal to customers. It was not feminine enough. Or was it because Charli in the British street language means cocaine? Which frame did Heineken use when starting the development of the name for the drink?

A year later, in April 2008, Jillz entered the market as Heineken's new women's drink.

CHARLI OR JILLZ?

EXERCIZE How would the development have gone if this frame had been tested at an early stage by potential customers?

There are many examples where examining frames is shortened or skipped. One frame is then automatically chosen and everything is viewed and interpreted based on that one frame. Only when implementing the solution does it become apparent that the end-user sees the problem from a completely different frame and the solution offers no answer for the problems, needs or wishes of the customer. It can already go wrong for a new product or service based on a wrong connotation of a word or product name.

What glasses do you wear?

What is your frame about breakdance? Is it just a macho thing among a bunch of hip-hop lovers? Or is it a unique dance form where it is all about personal style, originality and expression? The way you look at things, consciously or subconsciously, determines your follow-up actions. A group of young people coming to the town hall asking if there is space somewhere for them to be able to breakdance, will not be taken into consideration by a civil servant with the 'breakdance is a macho thing' frame, while a civil servant with the 'breakdance is a unique dance form' frame will look up the telephone number of a local dance school for them.

2.2.4 Process awareness

Within the design process you therefore always go through the cycle of design thinking, where you learn, every phase again, based on trial and error. Due to these continuous iterations, design thinking may not seem very efficient. Nevertheless, it is a very effective way of solving the underlying problem and developing a solution which is ultimately the best for the end-user

Process awareness

When you apply design thinking for the first time, it is sometimes difficult to estimate when a cycle of design thinking has been completed, a milestone is reached, or when 'good is good enough', so you can go on to the next step or phase of the design process. This gut feeling is also called process awareness. Whoever uses design thinking more often does not only learn to switch between the different steps and phases, but also gets a better feel for what to do, when to go back or when to continue.
Design thinker Lindsy Szilvasi puts it as follows: 'Design thinking is a mindful process. You don't want to get ahead of yourself, see what the next step should be, but you must be aware of where you are in the process at all times'. The same feeling must be developed for the deployment of the six fundamental attitudes in the cycle of design thinking. All fundamental attitudes are important during all phases of the design process.

2.3 The design process

The Design Council is the British government's independent advisory board for design and seen as the leading authority in the field of design in the broadest sense of the word. In 2005 the Design Council developed its own model for the design process, called the *double diamond*. The double diamond consists of four phases, divided into two 'diamonds'. In the first diamond it is all about discovering and defining the problem or challenge. At the transition point from the first to the second diamond, the problem you are going to tackle is outlined. This is done based on a *design brief*. In the second diamond it is about developing and delivering a result, the double diamond clearly shows the essence and structure of the design process. Figure 2.4 shows what the double diamond looks like.

Double diamond

In this book we will use the four phases of the double diamond as the basis for the design process, but we will enrich the process in our own way with the fundamental attitudes, the cycle of design thinking, phases, steps and milestones. The milestones make it clear where you are working towards to in the design process. As soon as you have reached a milestone, you continue on to the next phase. In Table 2.1 we translate the double diamond to the design process that we are using in this book.

FIGURE 2.4 Double diamond

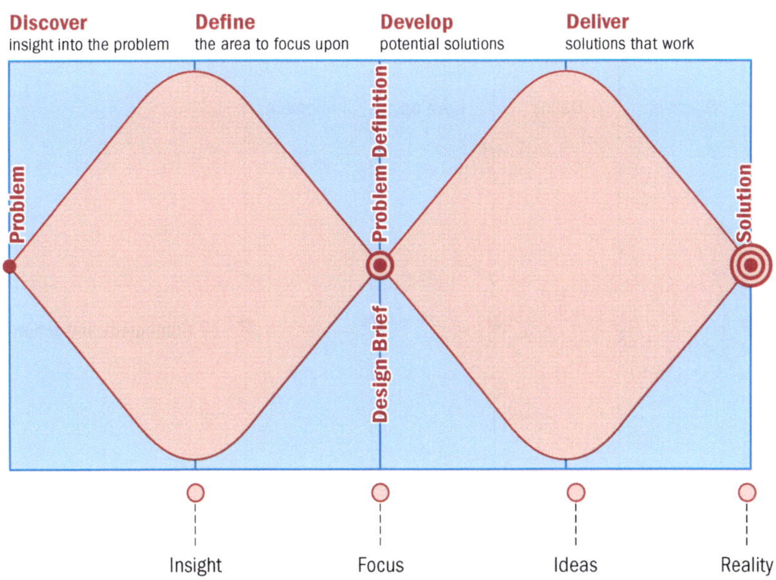

Source: Design Council (2005, adapted)

TABLE 2.1 Translation of the phases of the double diamond into the phases and milestones of the design process.

Double diamond	**Design process**	
	Phase	*Milestone*
Discover	Discovery phase	Insight into the problem situation
Define	Definition phase	Problem definition and solution area
Develop	Development phase	Tested possible solutions
Deliver	Implementation phase	Functioning solutions

Discovery phase

Definition phase

Development phase

Implementation phase

The discovery phase is a period of inspiration and insight, where you identify initial user needs, examine frames and come up with inceptive ideas about the problem. In the definition phase it is about organizing and prioritizing all the insights that you have discovered in the first phase and focusing on the insights that could possibly solve the problem. The main purpose of this step is to develop a creative instruction that frames the fundamental challenge; the so-called *design brief*. The development phase is centered on formulating, developing, concretizing, testing and refining potential solutions. Finally, you implement a functioning solution in the implementation phase. A preliminary answer to the problem has been given. Figure 2.5 shows the milestones per phase of the design process.

FIGURE 2.5 The design process in phases and milestones

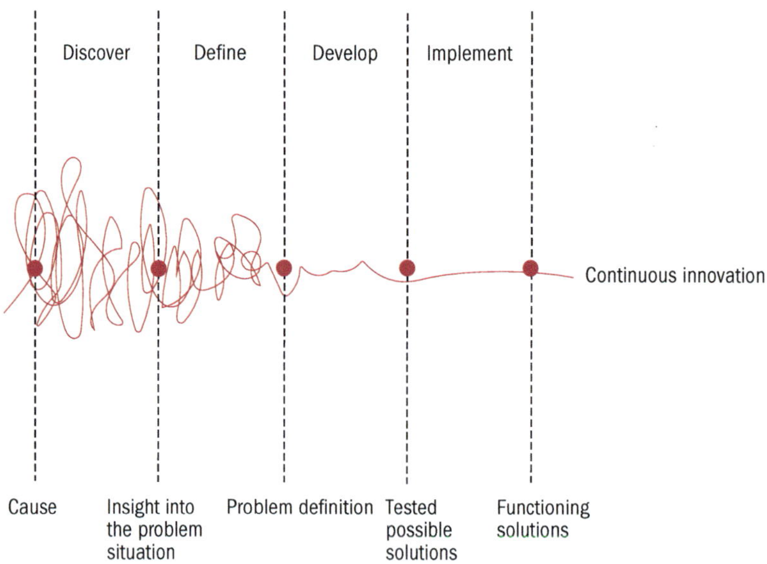

In the following sections we will discuss the four phases of the design process.

'One of my rules in consulting is simple: never solve the problem I am asked to solve. [...] Because, invariably, the problem I am asked to solve is not the real, fundamental, root problem.'

— Don Norman, author of *The Design of Everyday Things*

2.4 Discovery phase: loving the problem

There is always something that causes the start of a design proces. A cause can be a problem, an inspiring idea that requires elaboration, a question from customers to improve a service, etcetera. In this book we use the word 'problem' or 'design challenge' for all these variations in causes. The discovery phase is about investigating the cause and discovering the problem or design challenge.

Cause

FIGURE 2.6 The discovery phase

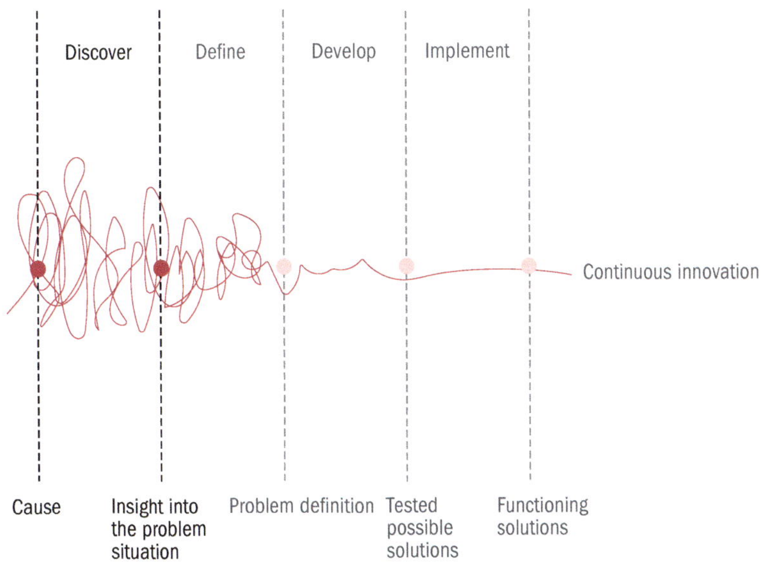

2.4.1 Problem as a source of inspiration

In a traditional and linear problem approach, the discovery phase, is usually shorter than when you tackle a problem with design thinking. There is pressure to quickly 'fix' the problem experienced by the client. This allows the client's problem definition to be seen as the 'fundamental' problem. The curious, inquisitive attitude is forgotten and the most obvious solution will be chosen. The result is a solution which may seem to work for a while, but ultimately does not meet the needs of the customer or user.

2

Reframing

What Volvo has done is special and an inspiring example [...]. Volvo involved designers, not car designers, in their thinking process. Not for designing the perfect car, but to stretch the frame of mind and form a vision for the future of driving. 'There are so many changes that we can basically cross out what a car looks like now,' says designer Miriam Van der Lubbe. Because who says a new car must be more perfect in the existing conditions? By reframing the issue (the work of artists and designers), new frameworks arise. This example shows that ambiguity (something that has different perspectives) belongs to the new age.

Design thinking assumes that the solution is 'in the problem'. If you do not know the problem, you cannot solve it either. The cycle of design thinking plays an important role in the discovery phase. The design team is not confined to that one frame for viewing the problem, but examines different frames. The trick is to see the problem as a source of inspiration, to ultimately 'loving the problem' and to convince the client to embrace the problem before wanting to come up with solutions. And if there is really no solution, is there actually a problem?

2.4.2 Action bias

Action bias

It seems to be in our DNA to immediately jump into the action mode when a problem occurs. This phenomenon is called action bias: human's natural tendency to take action because standing still is associated with indecisiveness. From the following examples of an action bias, it appears that this is not always the smartest move.

A goalkeeper almost always jumps to the left or right in a penalty kick, but staying in the middle offers the best chance of stopping it.

A doctor appears to prefer to start treatment than to admit that he or she cannot make a diagnosis at this time. A stock trader buys shares even if he has no knowledge of that specific market. An inexperienced police officer will want to intervene quickly in potentially dangerous situations while an experienced police officer knows that many situations resolve themselves (or do not escalate) by just waiting. According to author Robert Dobelli (2015) there is only one way to prevent that you jump into 'action bias mode': dare to take time out and wonder if you have enough insights to actually get started. Practice a week saying 'I don't know at the moment' if you get asked a question and discover how many questions actually solve themselves.

'Love the problem, not your solution.'

— Ash Maurya, founder and CEO Leanstack

In the discovery phase you must learn to sit on your hands and first thoroughly understand the reason for a design project. By actively researching, taking a dip in the past, you can find out why it was now that a design team was formed or hired.
In addition to gaining insights into the organization, the design team also focuses on how the user would experience the future and how he would view the client's problem: What problems does he experience? What needs and wishes does the customer have? In this first phase of the design process you look as broadly as possible into who is involved in the problem and identify these stakeholders. A design process can only be called successful if the result is accepted by the user and others with an interest in the problem or the solution. By not accepting and trying to solve the client's problem immediately, you first immerse yourself in all aspects of the problem, until you love the problem.

Standing still on the asphalt (1)

For decades, The Netherlands has been searching for solutions to their traffic congestion problem. Scientists, local interest groups, environmental activists and politicians all have their own idea about the solution or solution area. Scientists may be looking for a way to improve traffic flow, local interest groups would be looking at less emissions of particulate matter and environmental activists would look at CO2 emissions.
The ideas about the solution area seem contradictory and lead to discussions. It is precisely this that ensures that the problem is still there. If each of the named groups would let go of their solution area, then another process would start in which the problem is first thoroughly investigated by searching for the right questions. For example: is the problem in the number of kilometers of asphalt or in the (un)attractiveness of public transport? Is driving behavior perhaps not efficient or do people just work too far away from their homes? The fact that the solution area is unknown makes the outcome of the process uncertain. In a design process that is actually beneficial, the more solutions are suggested and the more stakeholders provide input, the more likely that solutions will be found that really work.

2.4.3 The problem paradox

Design thinking is often used for complex problems for which no *quick fix* can be found. An interesting factor with complicated and complex problems is that there usually is a contradiction hidden in the initial problem, which makes the problem so difficult to solve. One of the main questions in the discovery phase is: what makes this problem so difficult to solve?

Problem paradox

Kees Dorst (2015) refers to problems as paradoxes. A problem paradox is a contradiction within a complex problem, causing the initial problem not to be solved easily. Whoever tries to resolve a problem paradox gets stuck in circular reasoning: there is always a principle, law or standard that stands in the way of a logical measure. Dorst explains this on the basis of the 'fitting room problem'. In fitting rooms, a lot of clothing is stolen. Addressing this problem is difficult. Each logical corrective action (hanging cameras, supervising) is getting in the way and contradictory to the function of the fitting room (privacy for the customer) and the legal right to privacy. This is a problem paradox. Formulating this problem as a paradox helps to gain new insights. The problem paradox in the case of the fitting room, can be formulated as follows:

- Because people don't feel comfortable with the idea of being in the middle of the store in order to try clothing, stores have set up fitting rooms.
- Because a fitting room is meant to create privacy for the customer, you don't see what happens inside and it is a good place to hide something.
- Because a fitting room is a good place to hide, it is a good place to commit a theft.
- Because there are many thefts in a fitting room, measures have been taken to make the booths less private.
- Because the fitting rooms are less private, the customer no longer feels comfortable there.

We have come full circle; the paradox of the problem has been uncovered. In later phases, solutions can be sought via the design process, to break the circle that sustains the problem. The design team will therefore have to look for a suitable solution that eliminates the paradox or explores the limits thereof. Do you also notice that the curtains or fitting room doors have become shorter over time (which means that everyone sees your socks, but also what you have on the floor)? Or that more and more common spaces have been created around fitting rooms including banks for shop-goers (social control, despite experiencing privacy)?
Two more examples of problem paradoxes can be seen in Figure 2.7.

FIGURE 2.7 Problem paradoxes – traffic jams and app development

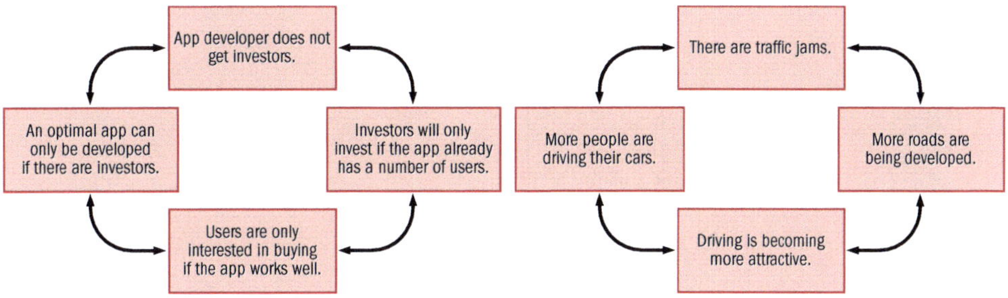

The paradox of the damaged bikes

Van Moof is a bike brand that is known for its high-tech bikes. The company sells its products through brand stores and online. When Van Moof started selling in the United States, a problem arose: many bikes that were shipped, arrived damaged at the customer's door. In the United States, the delivery services appeared to be less careful with cardboard packages: during transport, these boxes tore open, causing scratches to the bike frame or the wheels. Measures like new delivery partners, more robust cardboard boxes and warnings on the packaging did not solve the problem. What is going on here? The parcel deliverers seem to be stuck in the thought that bikes (in general) 'can take a beating'. A paradox! Every logical measure that was taken by Van Moof, such as warning 'Be careful, this is a fragile product!' is at odds with what delivery people think about bikes. Look on page 77 and find out what solution caused the damage to drop by 70 to 80 percent.

2.5 Definition phase: defining the problem

In the definition phase, all the insights that were discovered in the first phase, are ordered, prioritized and analyzed in such a way that there are a number of possibilities to go from there. This phase is all about focusing on the insights from the discovery phase that might lead to a solution of the problem.

FIGURE 2.8 The definition phase

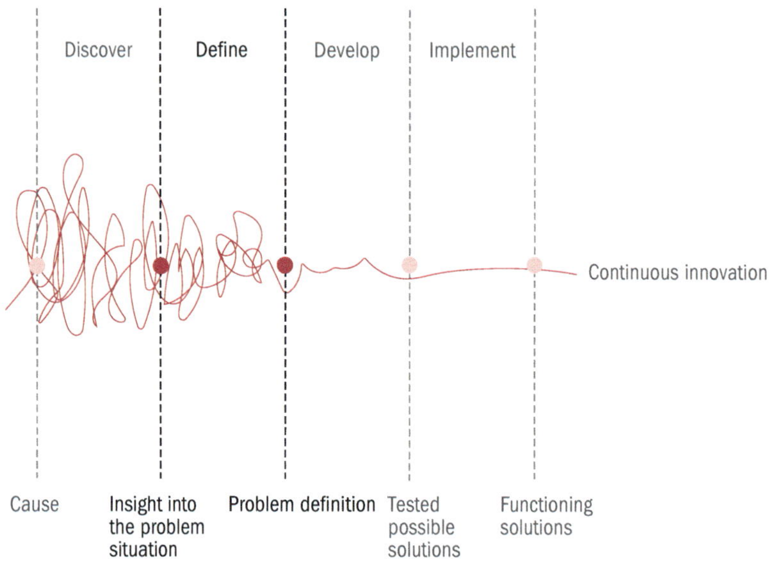

2.5.1 After diverging comes converging

The emphasis in the discovery phase is on diverging (viewing the problem from all angles and examining alternative explanations for the problem paradox), the definition phase focuses on narrowing down all the insights to a well-defined problem. This is done by converging: Which customers or people are we specifically targeting? Who are those people and how do they experience things exactly? What limits does the organization impose on us? What opportunities are there within these limits and what choices do we make about them? The definition phase is all about framing the problem so that it is workable within the context (and requirements) of the organization and the outside world. The definition phase ends in a problem definition, which is summarized in a so-called design brief.

In the definition phase, the fundamental attitude EMPATHIZE leads to profound research into the future user and his view of the world. Doorley, Holcomb, Klebahn, Segovia and Utley (2018) make concrete what you would do in your role as researcher:

- Observe users and their behavior in the context of their own lives and in the context of the problem space. You will be amazed how a problem can be viewed differently from the end-user's perspective. For example, when promoting carpooling (good for the environment, your wallet and the reduction of traffic jams), the fact that for many people, driving to and from work is the only 'me-time' that they have in a day is not thought about.

- Get involved by starting a dialogue. This is possible through long or short conversations. Prepare some interview questions as a guide, but leave the conversation to take its course. Realize that unexpected turns in the discussion bring about the most interesting insights. For this, it is necessary to ask open questions (who, what, where, when, how and why?) and by supplementary questioning. One answer is rarely enough.
- Look and listen: a combination of observing and engaging in conversation. Have a user physically portray what steps he goes through to do something that is relevant to the research and ask him to explain what he does, thinks, feels and experiences. For example, when developing a new packaging for a detergent, you may observe that the user has lost the cap. If you ask about it, you'll find out that this happens more often and that this person is really annoyed about that.

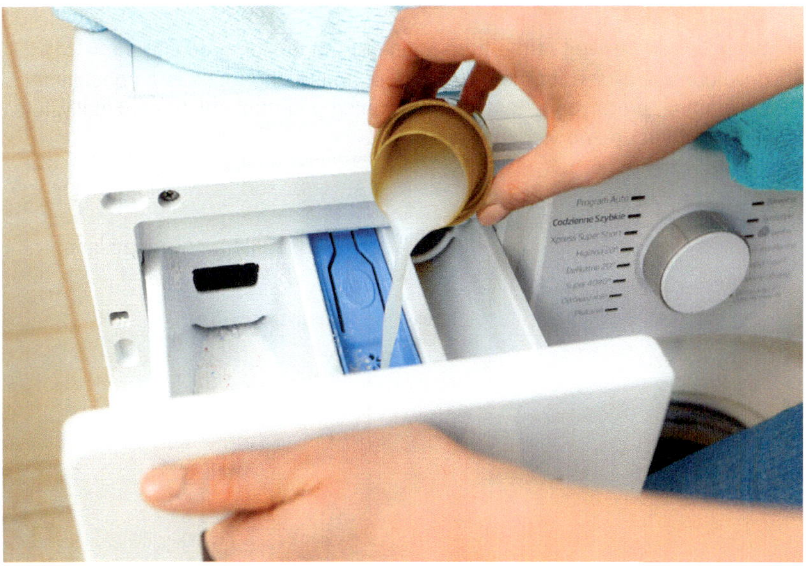

2.5.2 The design brief

A design brief summarizes the information from the first two phases. It is an organized representation of the information collected and includes a specific description of the (negative) situation that can be improved through adaptation, additions or innovation. A design brief describes cause-effect relationships that have been discovered and show the research results and figures that support the problem definition. The design brief therefore represents the core design challenge. It is the end of the first two phases and is the starting point for the continuation of the design process. So make sure it stimulates and inspires others. The design brief is the elaboration of one of the most important frames: it is a first suggestion for a framework in which solutions will be sought after and ensures a clear focus on generating solutions by the design team and others involved, something that was postponed until now.

Design brief

Good designers can create normalcy out of chaos

Jeff Veen, founder of design agency Adaptive Path, sees finding the common thread in a large number of ideas as one of the most important skills of a design thinker: 'Good designers can create normalcy out of chaos,' he says. There is not much professional literature to be found on this subject, which makes it seem as if finding the common thread is a magical step taking place in the designer's head. The link between the original ideas and the conclusions that designers draw is often not clearly visible. Jon Kolko investigated how this step works and observed that prioritizing, assessing and forging connections are important qualities that designers use to find the common thread. The magic is in getting a feel for when you can use which skill.

2.6 Development phase: working on solutions

We have now arrived at the third phase of the design process: the development phase. In the development phase, the focus is on designing, developing, prototyping, testing and refining potential solutions. Finally, it is time to think up and work out solutions!

FIGURE 2.9 The development phase

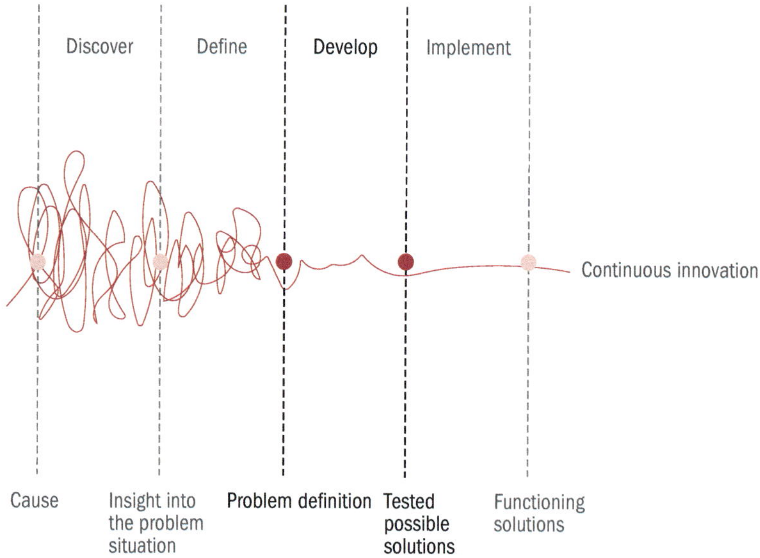

'An idea without form is merely potential.'

— Joost Backus, freethinker

2.6.1 Identify possible solution directions

Whereas in the previous two phases, the cycle of design thinking focused on the problem, it is now being used to find solutions. Because that is what it is all about from this moment on: researching ideas that lead to a possible solution for the defined problem which in practice actually works for the customer. First of all, possible directions for solutions are identified and tested. Then choices are made again to determine which directions will be developed into concepts and prototypes. This is a trial and error process, whereby the cycle of design thinking is repeated time and time again.

Stand still on the asphalt (2)

Let's take up the traffic congestion problem again and come up with some directions for solutions after we have investigated the cause on page 60. What if there are not too few highways, but too many ramps, causing traffic jams during short rain showers? Then there is a possible solution in allowing ramps to be used only in phases during heavy rain showers. What about a separate 'heavy rainstorm lane' such as the rush hour fast lanes that are increasingly found on motorways already?

And what if we let go of all this and link the congestion problem to our traditional way of working? Then it can stimulate flexible working hours, working from home or having schools start later (so that people do not all travel at the same time during peak times, which can also be a major contribution to the congestion problem).
The perspective of making public transport more attractive is also interesting. What if public transport is made more attractive due to tax reduction, shorter travel times or entertainment during travel time?

2.6.2 Coming up with solutions

Ideas are invented in all phases, but they always become more concrete in the development phase.
The following applies to generating ideas in this phase: come up with a lot of possible solutions within the set frame that have been established in the design brief. This step is about balancing between fluency and flexibility. On the one hand, you want a large number of solutions quickly (fluency). On the other hand, you want solutions that are truly different and distinct (flexibility). This makes sense because gathering many ideas that are a variant on the same theme is not constructive. In this step, there is a danger that people will get involved with a certain solution early on, and (unknowingly) leaving no alternatives to be considered. It is therefore important to combine speed and flexibility, so that an interesting mix of possible solutions is generated. In both cases it is about diverging: collecting as much as possible, different ideas. The probability that the solution to the problem is in there somewhere, significantly increases.

Super cheesy

Brainstorming originated in the New York advertising world of the 1950s. The idea behind brainstorming is that a high quantity of ideas leads to a high quality and usable idea. An example makes clear how brainstorming can be used within the design process.
You organize a brainstorming session about how to improve the promotion of cheese. You can randomly generate ideas by choosing a word from a book. Suppose you have chosen the word 'red'. The brainstorming session now starts with:
'How about a bright red packaging to attract attention?'
'Hey, strange, it already exists!' 'How about a new cheese that is red?'
'Or let's think conceptually. Red is connected to love. What do you think of a heart-shaped 'romantic' cheese?'
'When you start with brainstorming, you will soon find out that the possibilities are endless.'

We already know that after diverging comes converging. When converging, it is important to cluster and analyze the collected ideas and prioritize a number of concepts which can be worked out in prototypes. The terms 'idea' and 'concept' are often used interchangeably. We make a distinction between these in this book. An idea is a one-dimensional notion, standing on its own. The word comes from the Greek idio, which means 'individual'. A concept is multi-dimensional. Concept comes from the Latin con, meaning 'together,' and captum, a conjugation of the verb capio, which means 'capture' in the dual sense of the word: grasping and understanding. Several ideas can come together and be summarized in one concept.

Idea

Concept

The following example clarifies the difference between an idea and a concept. A design thinker is asked to make a hotel lobby more hospitable. Together with others he comes up with the following ideas: faster check-in options, a friendlier welcome, houseplants, counter removal, sitting next to the guest during check-in, offer a drink, create a restful moment, bring additional services that are offered to the attention of the guest. By combining the ideas 'houseplants', 'sit next to the guest during check-in' and 'creating a moment of rest', the concept of the homely hotel lobby emerges. Based on this concept, new ideas can be devised. What ideas can you come up with as a result of this concept? Keep converging and diverging!

An idea for the 'homely hotel lobby'

The concept of the 'homely hotel lobby'

Have all ideas been stolen?

In 2017 Prabir and Chakrabarti did scientific research on how experienced and less experienced designers come up with ideas. What proved to be the case? Designers almost always use an existing related idea from the past and adapt it to the current situation. In above-mentioned research, this principle is called *find & modify*: the existing idea becomes adapted to the current situation. Interestingly, most ideas come from your own past experiences and other people's past experiences.
Have you ever had an idea without using the principle of find & modify?

Which concepts would you ultimately work out in more concrete terms? And which not? Which would you present to users? This is a risky step in the process. Firstly, because there is a tendency to still save one's own idea or pick out the concept that you think will be 'fun' to work out. Secondly, because, in this step, there is a tendency not to involve the end-user in the process, as this would be too early. However, ask the future user to contribute

his thoughts in the first stage of the development phase; this provides a lot of information and helps to make objective choices.
At the end of this exercise, the remaining concepts that must be worked out into prototypes need to be prioritized. Only in this way can a well-founded choice be made for the solutions in the final step of the development phase which are then taken to the implementation phase.

Murder your darlings

Arthur Quiller-Couch gave a lecture on writing and literature at the University of Cambridge. In the book *On the Art of Writing* you will find the texts of his lectures. In one of his lectures about writing style he teaches his students to delete every irrelevant finery, extraneous ornamentation, in the text. He says: 'Whenever you feel an impulse to perpetrate a piece of exceptionally fine writing, obey it - whole-heartedly - and delete it before sending your manuscript to press. Murder your darlings.'

Sometimes it seems like an idea for a possible solution that has not been thought of before, is so fantastic that you prefer not to think about anything else and want to convert this brilliant idea into a product or service immediately. The challenge is to test the previously devised solutions against the set framework: If they do not match the design brief then they must be *killed* professionally.

2.6.3 Making solutions more truthful

Prototypes

Prototypes are used during all steps and phases of the design process. You continue to experiment and apply the cycle of design thinking, but you become more concrete as you progress. The testing of the prototypes developed in this phase is carried out by customers who will actually use the new product or service. By constantly involving end-users in the design process, not only does the design team receive (practical) feedback as to whether the solution works or does not work, but different concepts can also be compared. A good prototype provides the evidence that is needed to proceed unambiguously towards a certain solution. By 'unambiguous' we do not only mean that the members of the design team are in mutual agreement, but also the end-users and other stakeholders who are involved in the design process. A prototype is a powerful means of communication that allows the concept to be fully accepted by stakeholders.

When prototyping it is important to think about truthfulness. A prototype must be truthful and have a precision that matches the purpose of testing the concept. In design thinking this is called *fidelity*. There are different precision levels for prototypes:

Truthfulness

Fidelity

- *Low fidelity*: *quick* and *dirty* prototypes with low costs, which are rough versions and can be made quickly (within one hour). Examples: a sketch on a napkin or paper, one-hour prototype or storyboard.
- *Medium fidelity* prototypes: are still coarse, but more detailed and closer to the final solution. Time investment is half a day to a whole day. Examples: video prototype or role play.

'If a picture is worth a thousand words, a prototype is worth a thousand meetings.'
— Tom and David Kelley, IDEO

- *High fidelity* prototypes: these come very close to the final solution. They are very detailed and take a relatively long time to make (so that can be more than a day, but also a month). Examples of these are: minimal viable product (MVP, a minimal workable version of a product) and an experience prototype.

Boyle's Law: 'Never attend a meeting without a prototype.'
— Dennis Boyle, prototype builder at IDEO

After testing, the prototypes are improved, re-tested and in time, developed into real products or services that are to be implemented.

2

Lufthansa: next level prototyping

Lufthansa asked a design team to improve its service in business class on long-haul flights, 'beyond champagne and fancy food'. In the research phase, the design team flew business class with other airline companies. In this observation study, also called *participant observation*, the team gained an important insight: a personal connection between the crew and the passengers is the most important. In the design brief the following solution criteria was stated: how can the crew members provide more personal and individual service? Six new service concepts were devised from a collection of ideas; these were tested with a mock-up of an Airbus A380 on a 1:1 scale. This is next-level prototyping! For the development phase, the team chose this life-like prototype because small changes to the service can have huge consequences for flight attendants, ground staff and suppliers. An apparently simple extra action can lead to a lack of time and something as simple as a different spoon can lead to a lack of space in the always small storage spaces on the aircraft. Letting the staff experience how simple, but also how difficult it can be to apply some new concepts, proved crucial for improving the service in the business class of Lufthansa.

2.7 Implementation phase: towards functioning solutions in practice

The implementation phase is the final phase of the design process. Everything in this phase is about making the prototype (or prototypes) definitive and implementing a functioning solution that will be accepted by the customer as an answer to the problem.

2.7.1 Continue to apply design thinking

Translating the implementation phase into concrete steps and accompanying tools is quite difficult. Ricardo Martins (2016) made an analysis of different tools and discovered that of the 430 tools which he had reviewed, only 14 were about implementation. On the one hand, that is because in general, less attention is paid in design thinking to the implementation phase. Not because it is not interesting but because the steps in the implementation phase are very dependent on the context of the problem for which a solution has been devised: implementing an innovative solution for the congestion problem is completely different from putting an idea into practice for a more effective layout of a classroom. On the other hand, it could be that Martins found so few implementation tools, because iteration is normally continued during implementation. And for that you would use the tools from the first three phases of the design cycle again.

FIGURE 2.10 The implementation phase

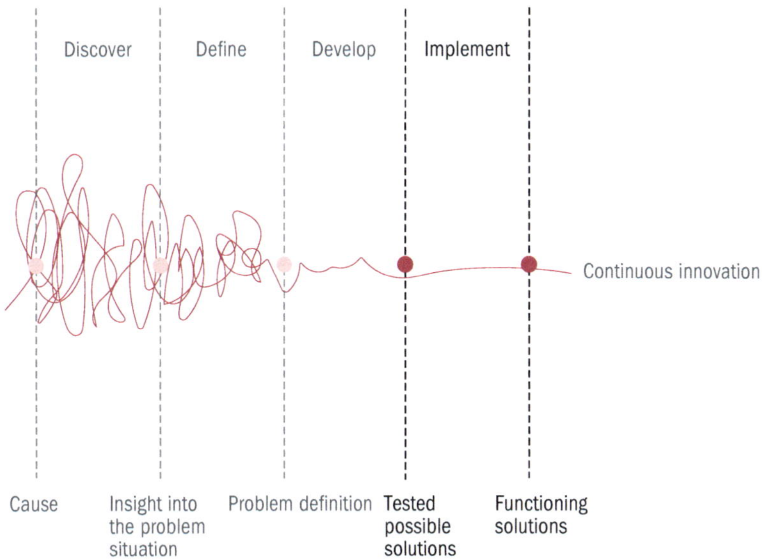

In practice, we see there is the greatest probability that the way of thinking and working in design thinking is abandoned in the implementation phase. Everyone is enthusiastic about the solutions found during the design process, but when implementing them in the organization, the pitfall is to implement the solutions 'as we have always done'. New solutions are then implemented in a linear manner, not tested any further and not continuously improved, with the risk that the solutions do not match the wishes and needs of the customer. To ensure that a solution fits the customer after the implementation and that it is continuously improved, it is important that the way of working and thinking in design thinking are broadly accepted in the organization. Within a design project this also happens by continuing to iterate during the implementation phase (continuing to test solutions with end-users and making last-minute adjustments).
Finally, a solution is gradually introduced into the organization and to its customers, so that people in the organization are not 'suddenly' confronted with changes and customers are not 'surprised' with a product they may not have asked for.

2.7.2 Launch and continuous innovation

(Parts of) solutions are sometimes already implemented in the development phase, before the full launch takes place. For example, this happens if an idea immediately benefits the customer after a quick test and can be implemented fast. The early implementation of (partial) solutions we call *quick wins*. Implementing the solution is a gradual process, in which employees and end-users are involved every step of the way when taking a new product or service into production and putting it on the market. That also holds true for testing the final optimally functioning prototype for the end-user. The design team keeps iterating with the cycle of design thinking,

Quick wins

for example by first introducing a solution to ten customers (and learning from them), then introducing it to a hundred customers (and learning from that) and only then implementing the solution for all customers.

Last minute adjustments

A municipality hired a design agency to enhance the visitor experience for the city hall which was under construction. When analyzing the future visitor flows, it was discovered that there was a mistake made in the design of the interior. The visitor would, upon arrival, bump into info columns for obtaining a ticket number for one of the counters. However, further investigation showed that 40% of the visitors did not really need a ticket and could go directly to the reception counter or to the upper floor. A logical and adequate solution would be to move the info columns fifteen meters so that the visitor would have more overview. The problem was that the implementation phase was already in full swing, the specification drawings were ready, the power lines already installed and that despite countless arguments against the columns, it was decided to place the info columns according to plan anyway.

The design agency then organized a meeting with project managers and the responsible foreman. At the town hall construction site, info columns made of cardboard were placed in the designated spots. To really experience this prototype, everyone involved got a role: someone was a father who came to declare the birth of his child, someone else had an appointment with the mayor, yet another had a question for the Permits Department and so on. Everyone walked through the city hall, looking for the right location. Within twenty minutes there was no doubt that the columns were situated 'really very strange'. It turned out that the project planning and the specifications could now be adjusted and the columns were placed fifteen meters away. Only by making a prototype could the deadlock be broken: the prototype communicated the message and clearly provided tangible evidence for what worked and what didn't.

Even after the full launch, the process of innovation does not stop. Continuous improvement of the solution and a continuous focus on the development of the original problem are essential parts of design thinking. An organization that sees a design project as being separate from the organization, will really notice during the implementation, that design thinking is a method that does not stop after implementation.

2.8 Using design thinking as a business strategy

Many managers see it as their main task to ensure consensus within the company. They prefer that nothing exciting happens. The can-do mentality of design thinking can clash with this mentality.
Within a design project, the design team can come up against the current standards of the organization, even if the project is working on solutions in isolation. Design thinking is embraced as a way of thinking and working, but the organizational structure, culture and internal processes are not (yet) tuned to the new way of thinking and working. The design team, which is focused on experimentation and making mistakes, is confronted with a culture in which 'first think long and hard' and 'make no mistakes' is preached.

Arts center De Kubus

Arts center De Kubus in Lelystad started an innovation team in 2017 in order to respond to the many social developments. One of the team's challenges was developing new services for the elderly. Myriam Cloosterman, advisor for cultural participation and leader of the innovation team, says: 'One of the first insights of the team was to find out the real question, as an important starting point when developing new services.' A design project was started in which various tools of design thinking were used: elderly citizens were asked to keep track of their daily activities in a diary on the basis of which they were interviewed. Soon after that, ideas for solutions were discussed with colleagues of the team and improved after feedback. 'We also made one-hour prototypes by portraying concepts for solutions in prototype flyers, with which we could test the possible services with the elderly in the streets.' Myriam Cloosterman found the process to be very successful. What she learned from this design project is that the service (the solution) consists of a combination of desirability (actual needs of the customer) and feasibility in the organization. From there, it is a matter of prioritizing the services that are ultimately implemented.

Messy beginning

Managers sometimes see the responsibility for an uncertain project as a (personal) risk. The manager would like to remove the sense of risk with an approved project plan, detailing exactly how the process is going to develop and what results can be expected. These managers find design thinking difficult. Design thinking has a messy beginning, because the problem as defined by the organization is not simply accepted by the design team. Because design thinking first examines the underlying problem, it is impossible to estimate exactly how the project will progress, let alone that there can be any promises about what the end result will be. Clients who do not want to start a design process if the outcome is not clear, use the same crooked reasoning as in job advertisements asking for a ‘very experienced graduate who knows what he or she can and wants’.

Throw the quotation in the trash?

If you are going to work as an external design thinker, be prepared for potential clients who cringe at the uncertainty about the results of a project. The client already expects answers in the quotation about the outcome: ‘Why do we pay for a process when the outcome is not certain?’.
A design thinker cannot provide the best solution for the given problem without first going through the design process. If the design thinker already knows the answers, it makes no sense to invest in the design process and the quotation can be trashed. The results would then already be available, but the chance that the right solution has been found for the end-user is small.

Uncertainty
Risk

Design thinking distinguishes between uncertainty and risk. A risk is about the probability that something will and can happen, expressed as a percentage. In case of uncertainty, you cannot know what is going to happen, let alone knowing something about its probability. Have you ever flown with someone with a fear of flying while it did not bother you? You probably immediately understood that the person was confusing uncertainty with risk. If you had calculated, for your anxious fellow passenger, how much more likely it would be to have a fatal car accident, this would, in all likelihood, have had no effect on that person.

If the design team has to deal with a manager or client who confuses the uncertain, messy start of the design process with running a risk, it helps to keep communicating about the natural flow of the design process. A team that has confidence in the design process cannot be tempted to come up with concrete solutions quickly, just to remove the sense of risk that the manager feels.

For the reader with fear of flying who is sensitive to facts

There are people who only feel secure when they are behind the wheel themselves. In a plane they cannot control the situation, which gives them a feeling of uncertainty. American statistics (reference year 2017) show that one person per 25 billion aircraft travelled kilometers is fatally injured. The same research suggests that this is already the case for road transport per 0.223 billion kilometers traveled by car. That means someone is about 112 times more likely to die in a car than in an airplane. Someone drives 30 kilometres from his house to Amsterdam Schiphol Airport and is flying to Barcelona. The risk of death during the half-hour car ride is 2.5 times higher than the two hours on the plane to Barcelona. However, the person with fear of flying estimates the risk of dying during a flight to be much higher. That's clearly not correct; he or she just feels more uncertain about the outcome.

2

2.9 Using design thinking as a business strategy

More and more organizations use design thinking as a business strategy. They do not see design thinking solely as a project approach, but as a way of thinking and working for the entire organization. That can be done by hiring design thinkers and by getting to know as many people as possible in the organization. Then design thinking becomes part of the business strategy. To promote the successful implementation of design thinking as a way of thinking and working in the organization, we have included some tips below:

- Ensure that management embraces design thinking as valuable business strategy and that managers know what design thinking is.
- Prevent that the support you get at the start of the design process gradually decreases. Keep management on board throughout the entire process by keeping them aware and active; and continue to communicate with them.
- Create awareness about design thinking. Not everyone needs to become a design thinker but let everyone get acquainted with the fundamental attitudes, process and tools. It is surprising how many people pick up on design thinking if they are involved in a design process.
- Stimulate skills and build self-confidence by letting everyone try design thinking, including managers.
- Start with a core design team that later supports and facilitates other projects. Create new design teams by putting specific teams together for each project (initially with representatives from the core design team).
- In the beginning, make sure that you are involving people from various teams and departments so that design thinking does not take place in a silo in the organization.

- Start with a number of small projects and small successes. If one of these projects is less successful, it does not immediately have disadvantages for the entire organization.
- Visually capture successful projects as persuasive evidence to implement design thinking throughout the organization.
- Design thinking is especially fun to do. Start looking for other organizations that are also involved in design thinking and exchange experiences. Learn from it and continue.

● www.emerce.nl

Design thinking among staff: 'Connect and say goodbye again'

Dutch organizations seem to have difficulty embracing 'design thinking'. This way of thinking is not new and internationally, there are already quite a few organizations applying it. In particular, keeping existing staff on board for this way of thinking often proves to be a challenge.
A buzzword or not, design thinking should be primarily seen as a way of thinking and a form of management, say the experts. The main things are customer focus and the idea that everyone in the organization can pitch in.
[...]
However, Dutch corporations do not appear to be massively involved in embracing this way of thinking. Gijs van Zon and Evert Hilhorst also notice this, both are employed at Freshheads. 'Designers are naturally accustomed to observing and iterating. But that is not part of the existing internal structures, 'says Van Zon. 'The most complex thing is that the people who are supposed to put it into practice for the longer term, tend to fall back on old behaviors. People are often blinded by standard ways of working.' [...]

Van Moof's innovative solution that enabled damages to delivered bicycles to be decreased by 70 to 80 percent.

Samsung

Lee Kun-Hee, the president of the Samsung Group, had been confronted with the lack of innovation in his company for some time. Staff members continued to stick to their old practices. In 1966 Lee Kun-Hee decided that he wanted to make a top brand out of Samsung. He reasoned that design 'is the ultimate battleground for global competition in the 21st century'. He realized that the company was lacking in design expertise and that this expertise would need to be developed despite the resistance that there was to innovate. To investigate the cultural, technological and economic trends, internal design teams were formed composed of engineers, marketers, researchers, musicians and writers. By choosing to develop the design culture with internal expertise instead of via external expertise, the entire organization gradually incorporated a design-oriented culture.

Ten years later, more than 1,600 designers were working at Samsung. How did the Samsung Group achieve this?

When a junior creative director got the idea to develop a phone without an external antenna, he initially did not focus on the appearance of the phone, as can perhaps be expected from a creative director. He knew his arguments about aesthetics would be hijacked by a technical counter argument. He decided to delve into the arguments from the technical development engineers.
This allowed the creative director to demonstrate that an internal antenna with a better range would be able to replace the traditional external antenna. This proved to be the decisive argument for engineers in developing (as it turned out) the phone that sold more than 10 million units.

When developing the Samsung Galaxy Note, a smartphone focused on making handwritten notes, it was hard convincing management of its usefulness. Such a combination of a tablet and a smartphone did not yet exist at Samsung or other suppliers. Comments like: 'You cannot hold the smartphone in your hand' and 'That won't sell' and jokes like: 'The only reason to buy it is because your face appears less large' were just a few of the criticisms. Everyone inside Samsung wanted to talk about innovation, but if something really innovative was presented, acceptance proved to be more difficult in practice. After a design review, where a high fidelity prototype was presented, management finally became convinced. The managers realized that they had looked at the initial idea in a very restrictive way.

Around 2003, researchers received an interesting insight from Samsung: regular household's TVs are more often turned off than turned on. The 'conventional wisdom' in producing television was always that image and sound quality were the most important. Next came the ease of use and the appearance of the television. Consequently, TVs all looked about the same and only differed in image and sound quality. But since TVs are frequently turned off, the appearance of the TV is just as important. Samsung designers moved the TV speakers from the sides to the inside of the unit, which caused some loss of sound quality but created a range of new options in terms of design possibilities. This was a landslide for the television manufacturer! Only after testing the market with a model that had concealed speakers, the organization was convinced that this development was the future for their TVs. The test turned out to be a hit and ensured that Samsung is now a leader in the field of design for televisions. See for example, The Serif, a television on legs and The Frame, a television that looks like a painting in the off position, which has a wooden frame.

Source: Harvard Business Review, 93 (9), 2015, by Young-jin Yoo and Kyungmook Kim

QUESTION 1
Which phases of the design process can you identify in this Samsung case story?

Identify them in the text: discovery phase, definition phase, development phase, implementation phase.

QUESTION 2
Do you also recognize the cycle of design thinking?

QUESTION 3
Which fundamental attitudes (or lack of it) did you detect in the article?

'Although everyone is for innovation, no one wants to change when we start talking about details.'

— Lee-Min-Hyouk, creative director Samsung Mobile

3 Design thinking is a project approach

In this chapter we will look at design thinking as a project approach. Based on a roadmap for a design project, we will let you practice design thinking. The roadmap includes the tools being used in a design project. This chapter is intended as a practical workbook for completing a design project and applying all the knowledge you have gained so far in this book.

In this chapter you will get the answers to questions such as:

- How does investigating the problem provide me with insight and inspiration?
- How can I prevent jumping to solutions too quickly?
- How do I gather ideas for solutions and make a choice from all the good ideas?
- How do I learn to test all possible solutions in practice and how do I make prototypes for this?
- How do I launch tested solutions on the market and what should I do once the solution is implemented?

KLM is investing heavily in design thinking

KLM Royal Dutch Airlines and Delft University of Technology signed a cooperation agreement called 'Design Doing at Royal Dutch Airlines'. The purpose of the collaboration is to create new products and optimize existing products and processes in a live KLM operational environment, so with real passengers at a real airport and in real airplanes. With this, KLM is making a considerable commitment to applying Design Thinking.
The faculty of Industrial Design Engineering at the TU Delft and KLM have been working together for some time to optimize the customer experience at KLM – among other things. The TU Delft offers and develops knowledge in the field of strategic design, while KLM offers the optimal test environment via its daily operations. Within this unique methodology, lessons from the design world, also called 'Design Thinking', can be used for the organization's strategy. To formally ratify the collaboration between the TU Delft and KLM for a longer period, the partnership called Design Doing was set up. Two of the TU Delft's affiliated PhD students worked together with other students on applying Design Principles in daily KLM operations.

Go to www.designthinkinginternational.noordhoff.nl and watch a supporting video about the article (viewing time is three minutes).

Source: www.travelpro.nl, March 16, 2017, by Arjen Lutgendorff

'What we want to achieve together is that projects go one step further.'

— Maïte Oonk

3.1 Introduction

You have gained knowledge on design thinking in the first two chapters. Now you will learn how to apply this knowledge in practice. We will give an example of a project approach, a roadmap for how a design project could be set up. The purpose of this chapter is not to present the standard roadmap for a design project but to imagine how a possible project would go. Every design project requires a different approach. The approach we propose in Figure 3.1 is only an option out of the many options that you have. We will walk you through the roadmap based on various exercises and develop them using the various tools that you will find in chapter 4.

Also use this chapter if you are going to work with design thinking for the first time. We will provide examples at the end of this chapter from other roadmaps to inspire you. The roadmaps are always based on the same structure, but are filled in differently each time. As soon as you've done a design project a few times, you'll find you can more easily estimate what, when and with which tools you can set up your design process.

In this chapter you will find questions and assignments to practice using the presented roadmap.

EXERCIZE

In this chapter you will find questions and assignments that you will need for your own design project when working with the presented roadmap.

YOUR PROJECT

In this chapter we refer to the tools in chapter 4 that you will need for your own design project when working with the presented roadmap.

TOOLS

FIGURE 3.1 Design thinking as a project approach

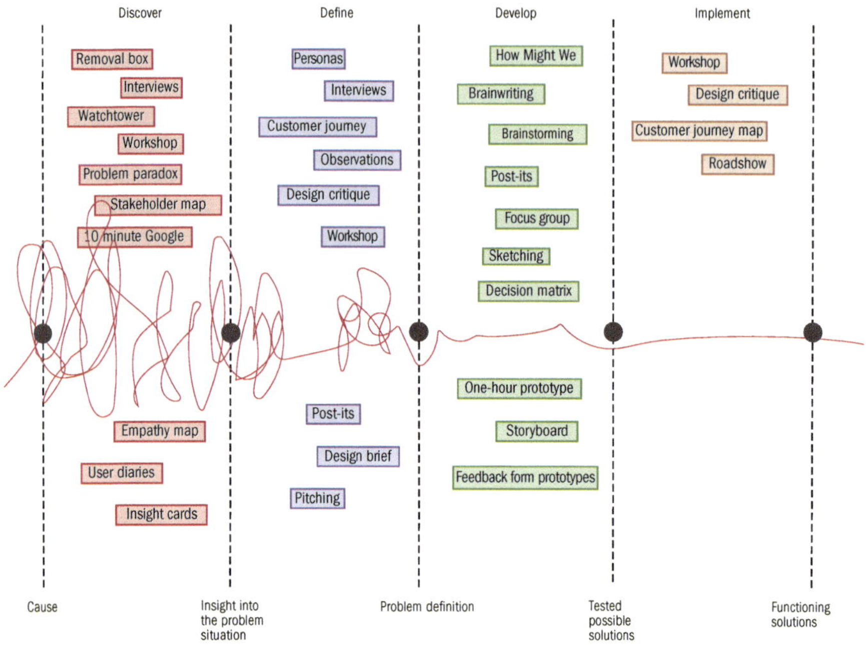

3.1.1 Preparing a roadmap

As mentioned earlier, design thinking does not work in a linear fashion and the design process has a messy and uncertain start. How do you plan a design project? Although within design thinking we always emphasize that you go through the iterative cycle, you cannot predict how many times a phase or step will go through iterations. However, the design project still has to be planned even if it is just something to hold on for the design team and others.

A roadmap provides a picture of what the design project will look like, which tools will be used and provides a planning estimate. A roadmap provides guidance for the design team and stakeholders. Compare the use of a roadmap with reading an actual road map in which you map out the route from A to B: you know that along the way you can make detours and stop earlier or later, but you still have a global picture of the route.

To be able to draw up a roadmap, the four phases of the design process will need to be completed in a very short time. Questions that are asked include:

- What is the lead time of the design project?
- Which steps can be distinguished within the phases?
- What milestones do we set?
- Which tools are used in which phase?
- How should the design team be put together?

'Building the bridge as you walk on it.'

— Robert E. Quinn, professor at the University of Michigan, School of Business

After each milestone, the design team reflects on the chosen path and updates the design process based on the advancing insights. A roadmap is therefore not definitive, but is constantly being fine-tuned as the project progresses. Do you need to convince a client about the roadmap that you have drawn up in the form of a quote? Also see the QUOTATION PROCESS.

Quick scan of the cause

Planning starts with performing a quick scan of the design process. Plan one or two hours for this and collect information about the cause for the design process and any other issues that you think matter. This is a mini version of the first phase of the design process. In a short time, you go through the cycle of design thinking a number of times until you have a basic understanding of the cause. Your first iterations! The point is that you gather just enough information so that you can proceed in planning the design process.

ASK YOURSELF THE FOLLOWING QUESTIONS:

- Has the project been done before in the past and is any information available?
- Which question was asked exactly (and to whom)?
- Who were involved?
- How does the client view the project? What does he or she expect from the results?
- Who are you going to solve the problem for?

YOUR PROJECT

Fuzzy front end

A novice design team often finds the uncertain and messy beginning of a design process difficult. What assignment are they saying 'yes' to? What exactly is expected of them? The uncertain start of the design process is also called the *fuzzy front end*. Fuzzy front end is the phenomenon that there are important decisions about the process that should be taken while everything is still unclear about the actual problem.

3.1.2 Putting together a design team

Another important aspect that must be arranged before the design process can really start is to put together the design team. A design team consists of people from different departments or organizations with different disciplines that jointly go through the design process to come to a solution for the defined problem.

Within organizations and project-based learning in higher education, people are often randomly put together in a team instead of being selected based on competencies and expertise. In such a case it will have to be examined whether the required expertise is nevertheless available and which roles can be fulfilled.
Ideally, a team is formed based on the expertise required, selected on the basis of in-depth knowledge and skills to ensure that there is cooperation across different disciplines. The team then consists of the so-called 'T-shaped persons'.

YOUR PROJECT

FILLING IN THE ROLES
Make team members jointly responsible for the design process by assigning roles at the start of the design project:

- The team captain is the cooperative foreman in the process who monitors and ensures that the team is making progress. In discussions, the team captain monitors the time schedule and the results that are to be achieved.
- The 'ambassador of the customer' constantly encourages the team to look at the process through the customer's eyes. By asking the right questions, the ambassador of the customer ensures that the end-user plays a central role in the whole design process.
- The challenger ensures that the team stays focused on the market opportunities and technical feasibility. With the proper questions, the challenger makes sure that the project team is really innovative in its solutions, without losing sight of reality.
- The connector ensures the correct connections are made between the team, internal organization and external stakeholders. The connector also searches constantly for the proper links with other project teams or other projects that are happening within the organization.
- A master of fun or experiment guru can also be added to the design team, or team members can be made responsible for monitoring the fundamental attitudes.

3.1.3 Getting started with design thinking for the first time?

Are you or your team starting with design thinking for the first time? Then first organize a kick-off. The kick-off is intended to provide information about the project with particular emphasis on how the design process works as a way of thinking and working. Take the time to address questions and concerns to remove any uncertainty about this different approach and always emphasize the added value of design thinking.

KICK-OFF

YOUR PROJECT

1 Prepare a presentation for your fellow students, in which you explain the design process as a way of thinking and working. Don't forget to pay attention to the cycle of design thinking and the six fundamental attitudes.
2 Invite an external expert to present a successful practical example in which design thinking was applied.
3 Want an example of how the author of this book tackles a kick off? Look at: www.designthinkinginternational.noordhoff.nl.

Imaginative space

Setting up a dedicated workspace is a good way to visualize design thinking as a way of thinking and as a way of working. A dedicated space for the design team ensures a greater tangibility of the design process: even if there are no tangible results yet, you can present the first research results and insights. An additional advantage is that early results will remain visible. This gives consistency to the design process.
The fundamental attitudes can also be exhibited in this physical space, such as: a visualization of the customer journey or personas that stimulate empathy; the presentation of all kinds of prototypes; a Playstation or a table tennis table for fun and team building; a visualization of the milestones of the design process that leaves no doubt as to where the design team is in the design process. A dedicated environment also ensures that the whole process can be communicated a lot better to customers, decision makers or other stakeholders that need to be involved in the design process.

3.2 Discovery phase: from cause to insight

In Chapter 2 we described how you must first deal with the problem, that you need to be inspired and not get into solution mode right away. We also described that in the fuzzy front end, there is quite a bit of uncertainty, while important decisions must already be made. Clarity about the cause for the design process is now essential. Understanding the cause for a design process is achieved by actively researching. This discovery phase ends in a joint insight into the problem situation.

FIGURE 3.2 Roadmap, phase 1

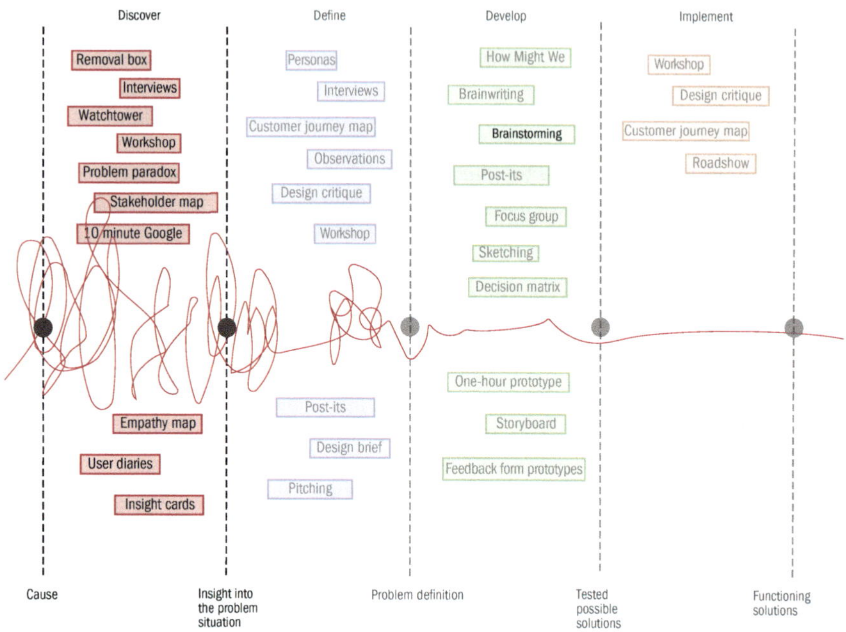

The following steps are worked out in this section:

1 take a dive in the past;
2 recognize the cause and paradox of the problem;
3 identify who is involved;
4 focus on the future user;
5 get a broad perspective on the problem situation.

3.2.1 Take a dive in the past

In practice, it often happens that the client presents the problem in such a way that it looks like it was created yesterday. In that case, pay attention and keep asking questions! The cause for the design process is often in the past. In many cases, previous attempts have already been made to solve the problem. Analyze why those solutions didn't improve the situation. Which steps have already been taken? Also investigate why certain steps were not followed: what would have happened if that one path had been taken? Also identify who was previously involved in the issue or problem and what role the client played. Has there perhaps even been a (design) team working on the problem before? Who wants to solve the problem and who is resisting? Also try to identify if there are limitations for the client that cannot be negotiated.

Those who take the time to dive into the past will be surprised by the amount of useful information that can be used in the design process. Be aware that you are only now seeing what others have not been able to see in the past. In hindsight, everyone is wise. Postpone your judgment by forcing yourself to view the problem from different angles and not to fill in the story by making assumptions. If there are any issues that are not completely clear, seek out a person who was involved and ask questions and summarize. Only then will you find out whether you have come to the right conclusions. No matter how difficult it is to not interpret using your personal frame of reference, visualizing the frame of reference of others will provide you with the information from the past.

BEHIND THE NEWS

EXERCIZE

1. Go to www.nytimes.com and choose a news headline that appeals to you.
2. Use the news headline as a fictional cause for a design project.
3. Look for answers to the following questions:
   - What has changed in recent times that is causing the problem to occur?
   - What will change in the near future?
   - What has already been done to solve the problem?
   - Why hasn't this helped in the past?
   - What has already been used to solve the problem (ideas, resources, expertise)?
   - What is the cause of the problem in time and money?
   - What are the possible consequences of the problem in time and money?
   - Are there already assumptions about possible solutions?
   - What is the culture within the organization where the problem occurs?

3

APPROACH FOR YOUR DESIGN PROJECT

TOOLS

A Collect background information with the REMOVAL BOX.
B Plan at least three INTERVIEWS with stakeholders from the past.
C Enter the WATCHTOWER.

Need more in-depth info? Consider using a FOCUS GROUP.

3.2.2 Recognize the cause and the paradox of the problem

After exploring the past, you focus on the problem that has arisen. Why are they asking a design team to solve the problem? What is it so annoying about the problem? What contradictions do you discover while exploring the problem? On page 188 we look in more detail at the problem paradox and we describe some examples to understand how you can get insight into a problem paradox by formulating cause-effect sentences (see also section 2.4.3). By playing with subsequent causes and consequences, you will come to so-called circular reasoning.
Use the circular reasoning you found to determine which additional research is needed to discover if you can get out of circle reasoning.

THE PARADOX OF RISING COSTS FOR HOME INSURANCE

EXERCIZE

Property insurance is becoming more and more expensive. Formulate, in four steps, a problem paradox for this fact.

FILL IN FORM 3.1 The paradox of rising costs of home insurance

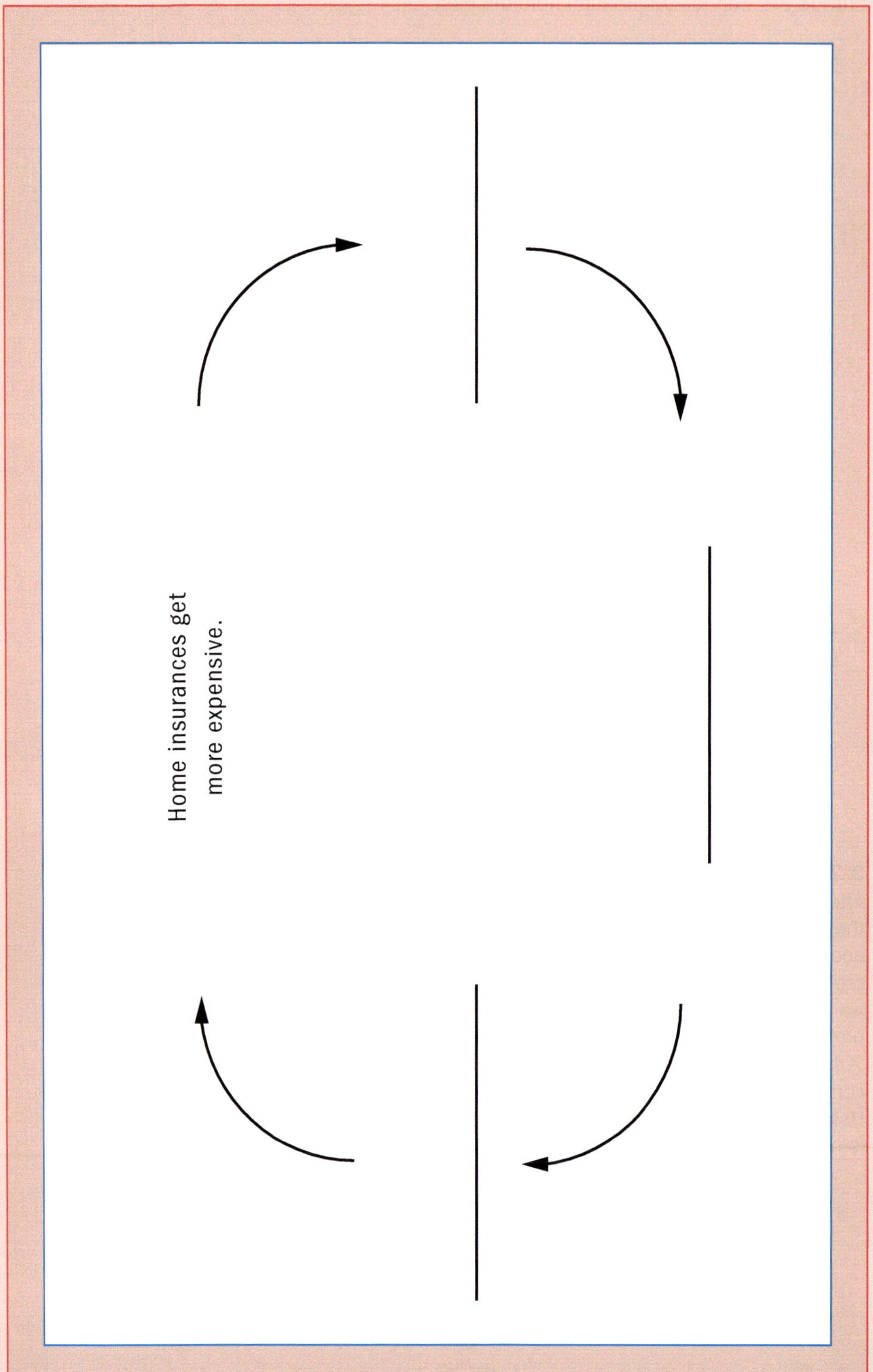

APPROACH FOR YOUR DESIGN PROJECT

TOOLS

A Organize a WORKSHOP for your design team to determine different problem paradoxes.
B Fill in a form for each PROBLEM PARADOX.
C Try to discover what sustains the problem and how the circle can be broken. Do not think in terms of solutions, but where there is 'space' to tackle the problem.

Need more in depth information? Work out different SCENARIOS to learn more about the problem.

3.2.3 Identify who is involved

In this step you put the data you found and the first ideas about the problem paradox(es) aside. That may sound strange, but distancing yourself from the problem history and the first idea about the problem helps you to think more freely. After all, every design process takes place in the wider context of the organization as a whole and of the world outside the organization. The temptation is to simplify the design project by disconnecting it from the confusing organization and the outside world that only make the project more complex. But just like every project, a design project takes place in the context of strategic, tactical and/or operational choices that all kinds of stakeholders have an opinion about. Ignoring the influence of these stakeholders is asking for difficulties. A stakeholder is a person or party who is interested in or can influence the problem and can therefore affect the design process.

There are three groups of stakeholders:

1. **internal** stakeholders (in the organization where the problem occurs);
2. **external** stakeholders (users, customers, shareholders, suppliers, financiers, press);
3. **interface** stakeholders (people in politics, local communities and education, whom you are dependent on because of laws and regulations).

In this step you investigate which interests and political (hidden) agendas there are and who can create resistance to (seriously) counteract the design process. Investing in a visual overview of the stakeholders and their demands and needs can be useful information for the course of the entire design process.

PRACTICING WITH THE SPIDER WEB

EXERCIZE

Suppose you are the president of a football club that has to merge with a football club a neighborhood away. With the use of design thinking, you want to ensure that the merger goes smoothly. Use fill in form 3.2 on the next page.

1 Write down the five most important stakeholders (first layer).
2 Write down five more stakeholders per stakeholder who can be involved as well (second layer).
3 Try to discover connections: who knows whom and how many more stakeholders can influence your design process?
4 Consider which stakeholders you should involve (even more) in the design process and how you want to do that.

FILL IN FORM 3.2 Practicing with the spiderweb

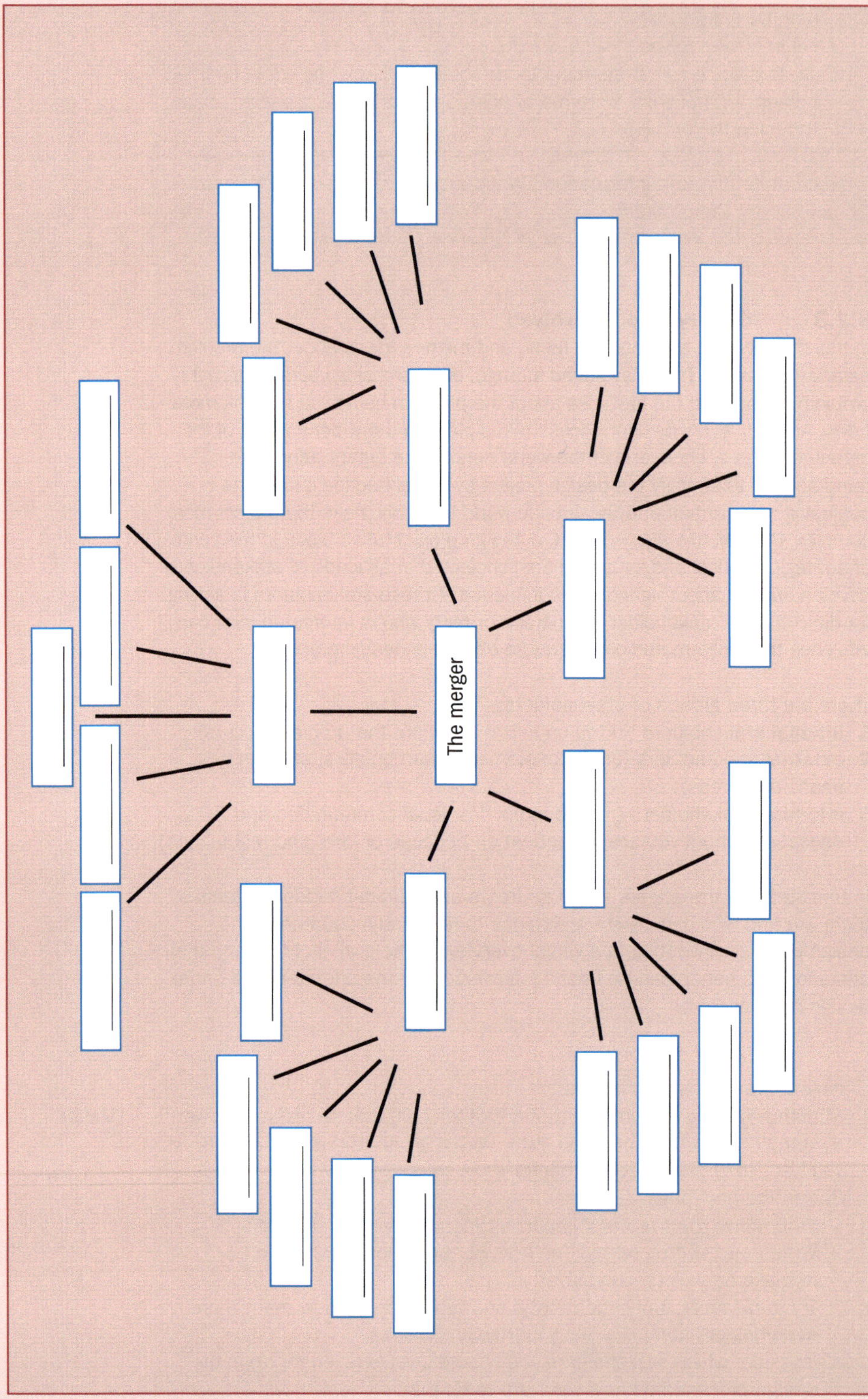

APPROACH FOR YOUR DESIGN PROJECT
Get started with the STAKEHOLDER MAP.
Are you short on time? Use the SPIDERWEB.

TOOLS

3.2.4 Focus on the future user

Now that you know who are involved in the problem, it is time to focus on who the future users of the possible solution will be. Now the fundamental attitude EMPATHIZE will clearly come in handy. The aim is to find out what the needs, motivation and experiences are of these future users. It is important that you constantly delve into this so that the final result of the design process actually meets the wishes and needs of the market. As you have read before in this book, you could for example, come to the conclusion via observation that people say they will do something but they actually don't do it in practice. Or that a user uses a workaround to avoid the problem, without being aware of it. By getting involved, you gain insight in the problem from the other's perspective.

3

INTERVIEWS are an essential part of the entire design process: it gives you a lot of information in different phases of the design process. In *Interviewing For Research, A Pocket Guide*, Andrew Travers (2013) names five steps to come to a research interview: recruit, prepare, conduct, document and synthesize. During the design process, interviewing customers, future users or stakeholders is a good way to empathize with others. If you already have used OBSERVATIONS, these will form a good basis for setting up your interviews. Interviewing can be used for insights from earlier observations but also works for testing frames, prototypes or for examining models.

Small children know how to get to the heart of a question. They continue to ask 'why?' until they have a satisfactory answer. The Japanese inventor and industrialist Sakichi Toyoda, founder of the company that later made passenger cars under the Toyota name, came up with the simple formula of *5 times why*: Ask the why question at least five times to find out what the root cause of a problem is. This also works if you want to know what is actually behind the needs of people and if you do not want to draw conclusions too quickly. By observing you know what you see but not why something happens. You will find out why by asking the why questions.

5 times why

WANT TO SEE THE REMBRANDT'S THE NIGHT WATCH (DE NACHTWACHT IN DUTCH) UP CLOSE?
'Do you want to see *The Night Watch* up close?' Young people who are asked this question will probably say 'yes'. Or not?
What do you see when you look at the image on the next page? What do you think is going on? Does this mean saying A and doing B?

EXERCIZE

3

Look again at the photo of *The Night Watch*. Suppose an interviewer keeps asking 'Why'? then the conversation could be very different:

1 'Why do you want to see *The Night Watch*?'
'I want to know more about *The Night Watch*.'
2 'Why do you want to know more about that?'
'I have to answer questions about it in my test.'
3 'Why do you have to do that in a test?'
'My teacher wants me to understand why Rembrandt painted the painting and which techniques he used. '

4 'Why does he want that?'
'Rembrandt is an important painter in Dutch art history.'
5 'Why is he an important painter?'
'Well, that's why I'm scrolling through the Rijksmuseum app.'

Do you now understand why the students in the photo are more interested in the app, a multimedia tour launched by the Rijksmuseum with background information about *The Night Watch*, than in the painting itself?

APPROACH FOR YOUR DESIGN PROJECT

TOOLS

A Use 10 MINUTE GOOGLE from the perspective of your most important users.
B Prepare different EMPATHY MAPS.
C Let some users keep track of USER DIARIES.

Need to dive deeper into the life of the user? Think of using OBSERVATIONS, QUESTIONNAIRES or SHADOWING.

3.2.5 Creating a broad perspective of the problem situation

In the previous four steps you have hopefully started to 'love your problem' and have collected enough information about the problem situation. It may well be that you already discovered solutions with apparently great potential during those four steps of the discovery phase. 'Eureka, problem solved!' you could even exclaim enthusiastically. Wait a while to draw this conclusion. Thinking in solutions narrows the perspective on the problem situation and slams the door on other solutions.
First it is time now to get an overall picture based on all the information collected and to outline and formulate insights. The insights provide a summary of all the collected information from the first four steps of the discovery phase.

A VISIT TO THE TOILET AT MADRID AIRPORT

EXERCIZE

Suppose a design team is asked to formulate proposals to improve the experience of the elderly at Madrid-Barajas International Airport. The client considers the following solutions:

- comfort (bench to rest on, lockers for hand luggage);
- additional services (offering deodorant);
- entertainment (music, giving information about destinations);
- hygiene (cleaning).

The design team first 'goes out on the street' and carries out a customer survey among elderly people at the airport through interviews, observation and inquiries. The research shows, among other things, that the elderly spend 5.5 times longer on the toilet at the airport than young people do: 22 minutes versus 4 minutes. The team decides to find the root cause of this fact by asking 'why questions': Why do the elderly stay in the toilet for so much longer?

If you were part of the design team:
1 What could be the follow-up questions? Imagine this from the perspective of an older person and answer them by your own or present them to your colleague or fellow student.
2 Which problem definition would you formulate based on your research?
3 Which solution possibilities would you pursue in order to achieve improvement, based on your answers?

If you have answered all questions, have you come to the same conclusion as the design team in Madrid?

Questioning by the design team revealed that the elderly stay on the toilet for so long, because this is the only place where they can hear the given flight information properly. Because of the bad sound system and the noise elsewhere at the airport, they could not understand spoken announcements and the visual flight information on the screens appears to be unreadable for the elderly. Through this insight you will get completely different solutions in looking for a more positive experience at the airport. Think of:
- silence rooms in the terminal;
- better sound system;
- better visual information about flights;
- an app with flight information.

Notice how you can be wrong if you immediately get into solution mode?

TOOLS

APPROACH FOR YOUR DESIGN PROJECT
A Fill in one or more INSIGHT CARDS to summarize the data you have obtained in the discovery phase.
B Do you want to test your findings? Use the content of the INSIGHT CARDS as the basis for an in-depth analysis in a FOCUS GROUP.

3.3 Definition phase: from insight to problem definition and solution area

The second phase, the definition phase, is about defining the problem: framing the problem in such a way that it can be worked on within the context (and requirements) of the organization and the outside world. The definition phase ends in a problem definition that is summarized in a design brief. In Chapter 2, we have defined the design brief as the marker in between 'loving the problem' and 'looking for solutions'.

FIGURE 3.3 Roadmap, phase 2

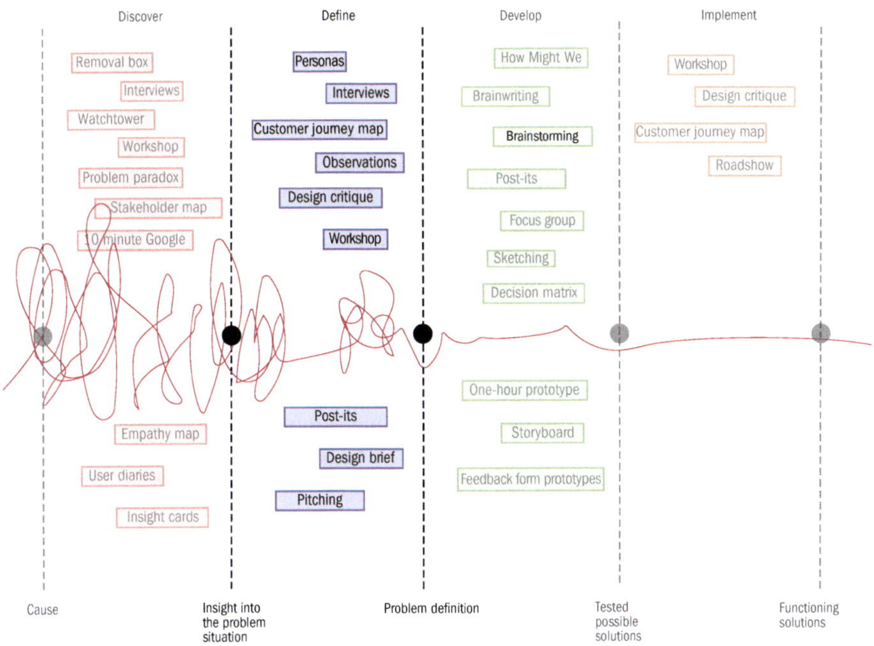

The following five steps are detailed in this section:

1 identify exactly who the user is;
2 find out what the user is experiencing;
3 determine the boundaries;
4 identify opportunities and make choices;
5 draft the design brief.

3.3.1 Identify exactly who the user is

In the first step, everything revolves around who the user is. In the discovery phase we already looked at who is involved in the problem as a future user. Based on the insights of the problem situation, you will now delve deeper into the future user or customer. The first question you must answer is: who actually is the user? You can discover this by developing PERSONAS. Personas are used in design thinking to visualize data from, for example, INTERVIEWS, QUESTIONNAIRES or other sources. A persona is worked out in detail and is visualized as a specific ‘person’ with his or her own unique character traits. Based on storytelling, the persona really comes to life. Using multiple personas helps the design team to identify with future users.

Personas are based on factual knowledge and insights into a broad range of users and represent a large group of users who in all likelihood, will behave in the same way. Personas are classified on the basis of character traits, so based on who the users are and not on, for example, demographics or purchase volume. Usually, four to five personas are developed that together constitute a good representation of the total user group. You can also prepare one or two non-personas; these are users who have nothing to do with the problem or solution.

When using personas, prevent stereotyping. Stereotyping is based on assumptions, prejudices and generalizations that are not based on collected facts and research. Also keep in mind that personas are meant to be an internal tool for the design team, not as a means to interact and communicate with users.

Customer profiles publicly disclosed

In 2018 Albert Heijn (the biggest Dutch supermarket chain) experienced that personas are for internal use only when their customer profiles were publicly disclosed. A 'premium' customer with a high spending pattern was depicted as a white male and a 'city budget' customer with a low spending pattern, as a colored woman. After this came out in the open, the company was accused of discrimination. 'We feel really sorry about this', spokesman Erik Lückers reacted in the Algemeen Dagblad newspaper. 'The selection of these images was unfortunate. There is no idea behind that, but now that people take offense, we will first take a critical look to see how this can be done differently.'

EXERCIZE

CLIENTS OF THE DUTCH ROADSIDE ASSISTANCE

You and your design team will make a client classification for the Dutch Roadside Assistance (ANWB):

1. Sit together in pairs and consider the basis on which the ANWB could classify its customers. Clichés are allowed, including ideas about age, gender, income and demography.
2. Google on 'ANWB persona' and check the video on www.designthinkinginternational.noordhoff.nl.
3. Compare your classification and that of the ANWB, and discuss the advantages and disadvantages of both formats.

TOOLS

APPROACH FOR YOUR DESIGN PROJECT

A Set up three PERSONAS.
B Test the prepared personas by means of INTERVIEWS.
Need to speed up or simplify? Create CUSTOMER PROFILES.

3.3.2 Retrieve what the user experiences

In the next step, everything revolves around what the user experiences. Now that you have obtained a picture of the different users based on the different personas, it is important to map out how these users experienced a problem. You do this by preparing a CUSTOMER JOURNEY MAP.

A customer journey, or customer journey map, is a visual representation of how a user experiences something. By visualizing the customer's experience, you find out where highs and lows are in the experience and which factors have influence on the client and the design team. Within the design process, the customer journey is deployed in the first two phases to gain insight into how a certain problem is experienced from the user's perspective. In the third and fourth phase of the design process it helps to imagine possible solutions for a client or have them tested by those involved. You can also compare the customer journey of the existing situation with that of the desired situation, to determine where improvements would be possible.

For a customer journey, think of the storyboard of a film in which the user's experience is portrayed step by step. Every detail is important and can be the key to finding a solution to the problem or improving a solution. Various experiences can be visualized during the customer journey. Think for example, about oral care. You can make a customer journey based on: 'How does someone experience brushing their teeth?' But also: 'How does a patient experience pulling a tooth?' Or 'How does a customer experience buying toothpaste online?' In the customer journey you look beyond the experience itself. You also look at what happens before and after the experience. What is someone doing an hour before, immediately after and an hour after brushing his teeth? With whom does the patient talk about pulling his tooth in the weeks before the procedure? What does the patient think when in the car on the way home? How does someone compare different websites, before proceeding with the purchase of toothpaste? With a customer journey, first pay attention to what someone does, thinks and feels. Afterwards, check where in the customer journey, there is interaction with the organization (or with the problem). The moments when the customer is in contact with the organization, are called touchpoints. Try to find out where there are gaps in the customer journey, so the organization can respond to this. A customer journey can be short (for example eight hours, during which brushing takes place), but there are also customer journeys that take a long time (for example, eight years, which includes selecting, taking out and using a dental insurance policy and switching to another insurance company).

There are different formats that you can use for drafting a customer journey. Two formats are included in the tool CUSTOMER JOURNEY MAP in chapter 4.

EXERCIZE

VISUALIZING A CUSTOMER JOURNEY

Google the keywords 'customer journey' or 'customer journey map'. Choose three visualizations that appeal to you and compare them:

1. Which parts match up in the three different customer journeys?
2. In which parts do the three customer journeys differ?
3. If you would make a customer journey based on 'ordering toothpaste online', which of the three visualizations would you use?

EXERCIZE

AN EXPERIENCE LIKE A MOVIE

A customer journey can be compared to a film that portrays the experience of the customer. Think of someone who gets his or her flat tire repaired at a bike shop. That is the main character of the 'movie'. What happened to the customer before the flat tire? And what did the customer do after the repair? The customer journey starts before the tire goes flat and ends after the tire has been repaired. The period before and after using a product or service, or preventing the problem, is what makes the customer journey interesting. Let every step pass by like a movie. Then make a storyboard of the movie in your head. Are you sure you have not skipped any steps in the process? Fill in the boxes on fill in form 3.3 on the next page, from the perspective of the main character. Two boxes have already been filled in for you, make sure you use the other fourteen.

FILL IN FORM 3.3 An experience like a movie

1	2	3	4
5 	6	7	8
9	10	11	12
13 	14	15	16

EXERCIZE

BUYING TOOTHPASTE ONLINE

1 Take the form for the customer journey map on page 139.
2 Distinguish the phases involved in purchasing toothpaste online.
3 Put the moment of purchase somewhere in the middle and fill in the rest of the activities afterwards. What activities does the user perform during which phase? What does the user do?
4 Consider and fill in what the user thinks and feels during each phase.
5 Consider whether and through which channels, the user comes into contact with the organization during the various phases. Enter these touchpoints on the form.
6 State the most important insights from the user's perspective, per phase, in 'highs' or 'lows'.
7 Do you already have ideas for possible solutions? Don't ignore them (despite that you are still loving your problem), but write them down for later.

Do you want to know how the author would prepare such a customer journey? Check it out on www.designthinkinginternational.noordhoff.nl.

TOOLS

APPROACH FOR YOUR DESIGN PROJECT

A Prepare a CUSTOMER JOURNEY MAP.
B Test the prepared customer journey by means of OBSERVATIONS.
If you have little time, then at least work out a STORYBOARD.

Need to test your findings? Test the prepared customer journey based on a series of INTERVIEWS or a FOCUS GROUP.

3.3.3 Determining the boundaries

Earlier in the discovery phase you explored the boundaries of the design project. Now it is time to test these boundaries in more concrete terms. By setting the boundaries, you determine the scope and status quo of the problem situation, and present this frame that you and your design team will work on, to those involved. You provide them with the insights concerning the problem in addition to what is achievable within the organization in the field of technology, finance, staffing or partnerships.

The design team must ask themselves what they will have to deal with in the last two phases of the design process. This will not be worked out in great detail yet, so not restrictive, since you are still defining the problem. You can organize a DESIGN CRITIQUE to determine the boundaries. The aim of the design critique is to take a critical look at current ideas about the problem situation. A design critique is often used to move a design process to the next step. By inviting people to criticize, the weak points of the status quo will come to the surface. In addition to the fact that this ensures further refinement (or rejection), it is a good way to involve (critical) stakeholders. Do not think in terms of functions or hierarchy, look especially for those who can make the best contribution. Think of: technicians, legal assistants, accountants, communication consultants, desk staff, etcetera. When determining who you involve in this step, you can use the STAKEHOLDER MAP again. A design critique increases the chance of success for the design process. If you are guiding the design critique, be aware of the following: you want those involved to be motivated to solve the problem, but at the same time you want them to see the barriers to problem solving and to communicate with you. This requires a safe and inviting atmosphere. The design team must be prepared to go back to the definition phase or even to the discovery phase, if criticism requires it.

Sunk cost fallacy

We know about the *Concorde effect* or sunk cost fallacy from economic psychology (Kahneman, among others). This fallacy means that you include the investment already made in your decision to continue or not. Although giving up is sometimes the best option, people often go through with something because it seems rationally 'logical'. This happened, for example, during the development of the Concorde, a super plane that was developed in the last century. The French and British governments kept investing in the plane, even when they already knew that using the plane would never be feasible. Do you sometimes get caught in a sunk cost fallacy? For example, you stayed in the cinema watching a movie you did not enjoy because it would be a waste of your money spent on the ticket? Don't fall for the Concorde effect and remember that you have already paid and staying will cost you precious time on top of the money already spent.

APPROACH FOR YOUR DESIGN PROJECT

Organize a DESIGN CRITIQUE with the most important stakeholders, in which the insights into the problem are tested against what is feasible for the organization in the area of:

- technology;
- finance;
- staff;
- partnerships.

TOOLS

3.3.4 Identify opportunities and make choices

In the previous step, the boundary determination, you have already started to converge. The identification of opportunities is about converging even further; making a choice on which you will focus: Which frame will you use in the development phase?

3

For this, you can use POST-ITS or the HIGHLIGHTER. By writing all the insights you have gathered on post-its or printing all information gathered and highlighting the most important findings, you can bring the large amount of information back to a core. The advantage of post-its is that you can physically move the information around and make connections a lot faster between groups and clusters. Highlighting mainly has a time advantage and is more focused on prioritizing insights according to importance. After clustering, the large amount of insights that you have gathered can be visualized by making a MAGAZINE COVER, as an alternative to a summary in words.

EXERCIZE

BUNDLING SOLUTIONS FOR THE TRAFFIC JAM PROBLEM

Figure 3.4 shows twelve insights on post-its, written in simplified form, of possible causes of traffic jams:

1. Use POST-ITS to cluster the insights.
2. Compare your layout with that of others and explain how you came to this classification.

FIGURE 3.4 Possible causes of traffic jams

APPROACH FOR YOUR DESIGN PROJECT

Organize a WORKSHOP based on POST-ITS and bundle the various ideas, insights and information about the issue.
Do you want to visualize the results? Use the MAGAZINE COVER.

TOOLS

3

3.3.5 Preparing the design brief

A design brief marks the end of the first two phases and launches the last two phases of the design process. The design brief must be a stimulant to continue to move on with the design process in a productive manner. A design brief marks the transition between 'loving the problem' and 'looking for solutions'. The most important goal of this phase is ensuring that you get started with tackling the 'right' problem. The design brief represents the core design challenge.

Design thinking takes the organizational perspective into account, but looks at the problem primarily from the customer perspective. The design brief therefore translates the insights from the first two phases of the design process to the user perspective, so that in the last two phases, there is continued focus on the user. When drawing up the design brief, do not be tempted to formulate the answers to the problem of the organization.

TOOLS

APPROACH FOR YOUR DESIGN PROJECT

A Prepare a DESIGN BRIEF.

B Actively involve those participating in the chosen direction. Present the ideas from the design brief based on PITCHING. Do you want to meet those involved? Organize a ROADSHOW.

3.4 Development phase: from solution area to solutions

It is finally time to come up with solutions and develop them. Because that is what this phase is all about: coming up with, and researching ideas which are a possible solution for the defined problem that will actually work for the customer in practice. First of all, solutions will need to be identified and tested; choices have to be made again for the solutions that need to be developed into concepts and prototypes. The development phase ends with a set of tested solutions to the problem.

FIGURE 3.5 Roadmap, phase 3

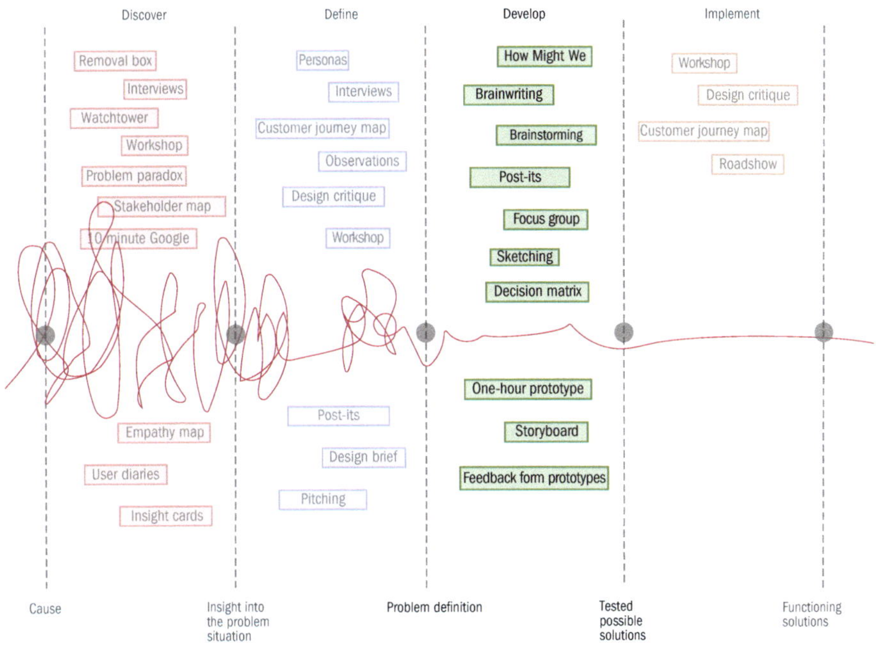

3

The following three steps are elaborated in this section:
1 generate ideas for possible solutions;
2 further develop ideas into concepts for solutions;
3 concrete testing of solutions.

3.4.1 Generate ideas for possible solutions

Whereas the definition phase ended in converging with the design brief as a tangible result, the development phase again starts with diverging: generating ideas for solutions. The focus is on coming up with many possible solutions for the given design challenge as set out in the design brief. On page 66, we already mentioned that it is about learning to balance fluency (coming up with many possible solutions in a short time) and flexibility (coming up with distinctive possible solutions). In this step the emphasis is on raw, crazy, strange or super simple and logical possibilities, that are not yet elaborated or conceived. Everything is possible in this step and everything is okay in this step. There is therefore no format or list available of characteristics that should form an idea. As a design team try to formulate ideas not only in words, but also think of the fundamental attitude IMAGINE and work with sketches or make a mood board.

Don't forget who you are doing it for

In 2017, So and Joo conducted research into the use of personas and its effect on creativity. Two groups of students were asked to brainstorm. The first group received a case: 'Bernd is 40 years old and manager at Siemens. He is married and has two children. Bernd works 45 to 55 hours a week. He leads a team of eight people. He gets stressed out by the frequent deadlines and has no time for his hobbies, sports and music.' The second group was only asked: 'How can you help an employee to experience less stress at his/her work?' Both groups received the same task: 'Please write down as many ideas as possible how to solve this problem.'

The outcome convincingly demonstrated that the use of personas is effective: the first group generated significantly more original ideas. The scientists concluded that by using personas, students were able to associate more freely. They compared the case with situations they knew themselves. Through these associations, additional information about the situation emerged which led to more original ideas.

To develop ideas, this roadmap starts with formulating *How Might We* questions. With HOW MIGHT WE you translate the insights from the design brief into different action oriented ones with different perspectives:

1 Emphasize what is positive.
2 Make a comparison with something else.
3 Split up the insight into parts.
4 Remove the negative aspect of the insight.
5 Question assumptions.
6 Identify unexpected options.

After formulating How Might We questions you start looking for answers to these questions. You can do this individually or in a group. Do you work in a group? Then you can do BRAINSTORMING or use BRAINWRITING. Brainwriting is the silent variant of brainstorming. Instead of calling out ideas, they are written down. When writing down ideas everyone gets a chance and ideas are formulated more thoughtfully. Brainwriting aims for the same cumulative effect as brainstorming (ideas get bigger by reaction to reaction), but then in silence.

> 'As sexy as brainstorming is, with people popping ideas like champagne corks, what actually happens is when one person is talking: you're not thinking of your own ideas.'
>
> — Leigh Thompson, professor at Kellogg School of Management

REAL LIFE EXPERIMENT

You will now investigate to what extent the results of a brainstorming session and a brainwriting session differ.

1 Create two groups with your classmates. Make sure there are two different spaces to work in.
2 Both groups come up with ideas for 'new flavors of soda'. The one group starts with brainstorming, the other with brainwriting.
3 Analyze and exchange the ideas devised in the different groups.

EXERCIZE

APPROACH FOR YOUR DESIGN PROJECT

A First formulate a number of How Might We questions individually.
B Then formulate a number of How Might We questions again with your design team.
C Use BRAINWRITING to come up with answers to the How Might We questions.
D Work out the ideas that have arisen using BRAINSTORMING.
E Bundle the ideas via POST-ITS.

TOOLS

3.4.2 Further development into concepts for solutions

After looking for answers to the How Might We questions you have combined the best of different ideas, dumped non-working ideas and bundled the result into a pair of robustly designed, multi-dimensional concepts. You can easily compare a concept with a crowded wardrobe from which you combine clothes (ideas) until you have created the best outfits (concepts). On page 66 we explain the difference more detailed between ideas and concepts. The conceived and clustered ideas from the previous step are now slowly taking shape. This is a good time to do additional research by involving end-users. What additional research is needed to get the clusters to convert to concepts? For example, you can do more INTERVIEWS, OBSERVATIONS or FOCUS GROUPS.

Again, you have to choose which concepts to continue with. Think in advance about the basis on which you are going to prioritize: once the concepts are on the table, it is more difficult to determine the basis on which to prioritize. Check, for example, the matters that are involved in step 3 of the definition phase (scope). Determine the guidelines on which you will evaluate concepts. A DECISION MATRIX helps you to prioritize the insights, problems or solutions found. This is a table in which factors are compared with choices or options, to be converted into scores. Filling in a decision matrix helps making the right choice based on a factual comparison, instead of based on a feeling or intuition. Consult with others about the scores and determine together with them which concepts you end up with. Do you work for a client? Make sure you actively involve him or her in this important step.

TOOLS

APPROACH FOR YOUR DESIGN PROJECT

A Organize an assessment with a FOCUS GROUP based on several SKETCHES of the developed concepts.
B Fill in a DECISION MATRIX and make a choice for the concepts that go to the next step.
C If you have a client: involve him/her in a WORKSHOP.

Do you want to visualize the results even more and/or test more extensively with users? Transform the concepts into SCENARIOS, a CUSTOMER JOURNEY MAP or CHARACTER PROFILE and present them to future users via INTERVIEWS. An alternative to making a choice from the developed concepts is the COCD BOX.

3.4.3 Concrete testing of solutions

We already have referred to testing concrete, detailed concepts as prototyping. Prototyping happens throughout the entire design process, as we learned earlier when we discussed the fundamental attitudes IMAGINE and EXPERIMENT.

How many people do I have to involve in testing the solution?

How many customers or users do you involve in testing developed solutions? Don Norman (2013) suggests being pragmatic: 'Five is usually enough to give major findings.' Fine tune your solution based on these insights first. Do you want to involve more people from your target group right now? Then ask five other people and make them test the adapted solution. Continue until you have the number of people involved that you deem desirable. It is better to have five variants of a solution tested by five, than one variant by twenty-five people. By the end of the design process, a large number of people will have worked on finding the solution.

TOOLS

APPROACH FOR YOUR DESIGN PROJECT

A Make low fidelity prototypes of the solution(s) based on a ONE-HOUR PROTOTYPE.
B Imagine the solution(s) from the user's perspective using a STORYBOARD.
C Test the one-hour prototype and the storyboard in a FOCUS GROUP.
D Use the FEEDBACK FORM PROTOTYPES to record your findings about the solution you tested.

More extensive testing? Organize a DESKTOP WALKTHROUGH or ROLE-PLAY.

3.5 Implementation phase: putting solutions into practice

The tested solutions are implemented in the fourth and final phase. The implementation phase results in fully functioning solutions.

You continue to iterate during the implementation phase: a continuous search for the best ideas by making them concrete, testing and observing what works. You want the functioning solutions from the previous phase, the development phase, to also function in the day-to-day operations of the organization and that the solution seamlessly meets the wishes and needs of the customer. Only if the solution is accepted by the user, can there be a successful implementation.

FIGURE 3.6 Roadmap, phase 4

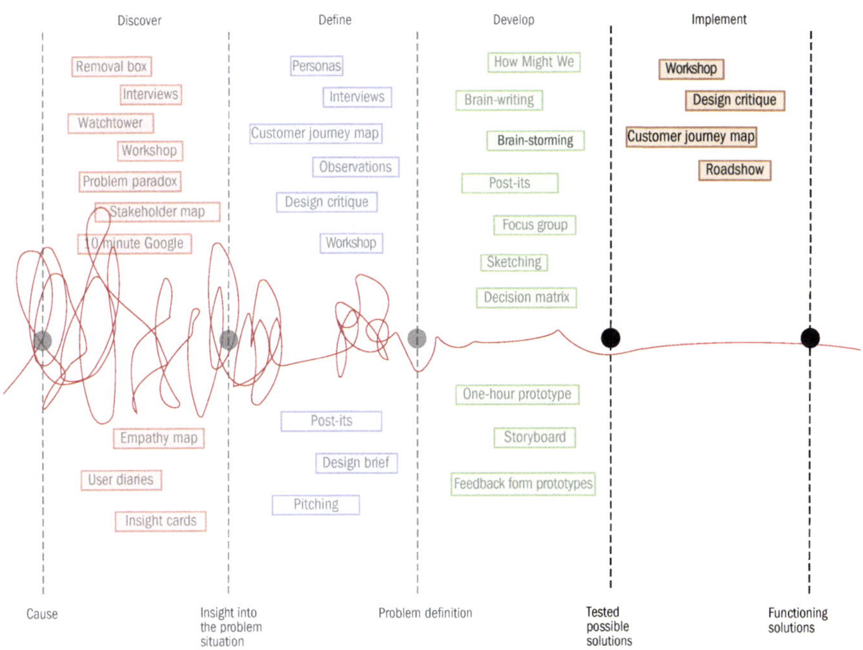

The following three steps are detailed in this section:

1. going back in time;
2. continue testing in day-to-day operations;
3. obtaining approval and launch.

3.5.1 Going back in time

Whereas planning the design process started with a quick scan as a forerunner of the design process, the implementation phase starts with reviewing all the work that has already neem done in the previous phases. Looking back is important because in the development phase the ideas, concepts and prototypes that have been devised and preliminarily tested and the information from the discovery and definition phases may have unknowingly disappeared to the background.

YOUR PROJECT

BACK IN TIME

Take one to two hours to get together with the design team to review all of the following steps:

- Take the results of the REMOVAL BOX: Have we really approached the problem differently than others who have already been dealing with the problem?
- Go through the formulated PROBLEM PARADOXES: aren't we (also or again) caught up in circular reasoning?
- Grab the created STAKEHOLDER MAP:
  - Do we actually have those participating who need to be involved in the process?
  - What role do these stakeholders play in the implementation phase?
  - Do all involved support the implemented solution(s)?
- Grab the developed PERSONAS and the CUSTOMER JOURNEY MAP: have the personas and the customer journey been improved based on the most recent solutions?
- Has all feedback from the DESIGN CRITIQUE really been included in the development phase?
- Which ideas, concepts and prototypes have been discarded earlier, but now provide interesting insights?
- Let the members of the design team also share personal anecdotes:
  - What are their most important 'aha experiences'?
  - Which brainstorming session or team meeting will they not soon forget?
  - What insights have they gained through their experience? Was there a real eye opener?
  - When did they laugh enormously during the design process?
  - Which experiment was the biggest failure?

TOOLS

APPROACH FOR YOUR DESIGN PROJECT

Organize a WORKSHOP with the design team. Ask the individual team members to formulate answers to the questions from the previous assignment.

3.5.2 Continue testing in day-to-day operations

As described in the introduction, iteration simply continues at this stage. After going back in time, you start performing the final testing and then proceed seamlessly to the next step: obtaining approval and launch. In the previous step you have looked to the past so be sure not to overlook any important insights about the implementation phase.

In the final testing you must check whether the organization where the solutions are to be implemented, is ready for implementation. Think about:

- Feasibility. Is the solution still possible functionally, financially, technically and in terms of capacity? Does the solution make the organization stronger?
- Viability. Is the business case behind the solutions well thought out? Is it possible to earn sustainable money with it?
- Desirability. Does the end-user want this solution?

Parallel to continuing testing, you work on the implementation plan for the solution.

APPROACH FOR YOUR DESIGN PROJECT

A Organize a DESIGN CRITIQUE again with stakeholders in the organization.
B Draw up a CUSTOMER JOURNEY MAP for the solution and test it with a few users.

TOOLS

3.5.3 Obtaining approval and launch

Launching does not mean the solution is implemented 'all of a sudden' in the organization and that customers are surprised by the new product or service. In most cases, the solution has already been implemented with ten customers (or branches of the organization). After some adjustments, the implementation may have been scaled up to the following 50, 1000 or 5000 customers (or branches). By gradually implementing, the distinction between the development phase and implementation phase can become less visible.

You need to do the following in this step:

- Consider who (formally and informally) must give the final approval and involve these people in the launch.
- Continue to use this step for last-minute improvements to solutions: iteration is and remains a continuous process.
- Organize an evaluation meeting with the most important stakeholders and the client in which not only the solution, but also the entire design process is evaluated. Does the organization want to apply design thinking more often in problem solving?

APPROACH FOR YOUR DESIGN PROJECT

Organize a ROADSHOW with stakeholders in the organization to evaluate a design project.

TOOLS

3.6 Other roadmaps

Every design project requires a different approach. The approach you have been going through in this chapter is just one of many you could use. Below and on the next page you will find a number of alternative roadmaps that you can use for further practice. Ideally put together a roadmap yourself, but one of the roadmaps can also be adapted and serve as a planning for the next project for which you want to apply design thinking.

FIGURE 3.7 Variation on the presented roadmap

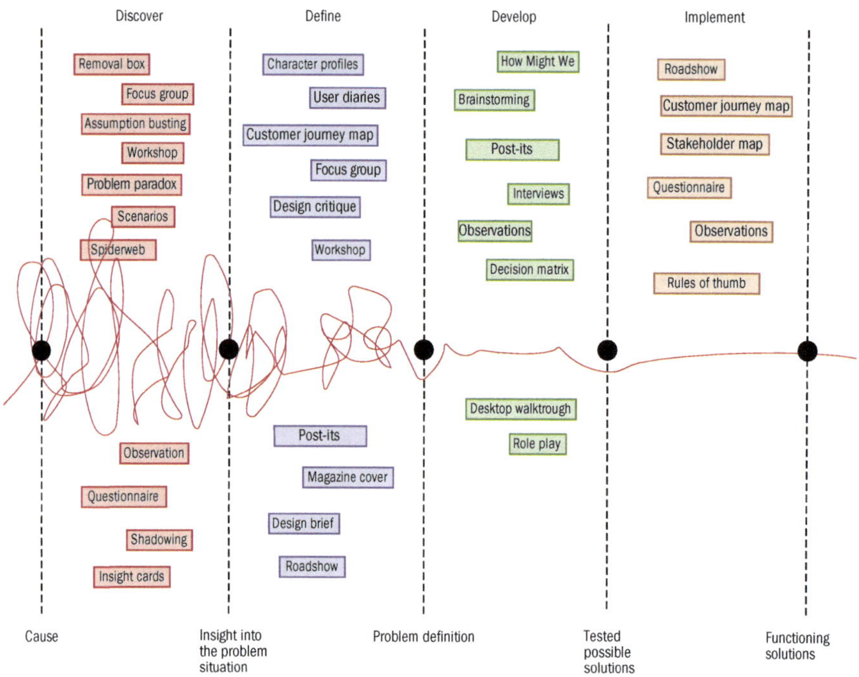

FIGURE 3.8 One week roadmap

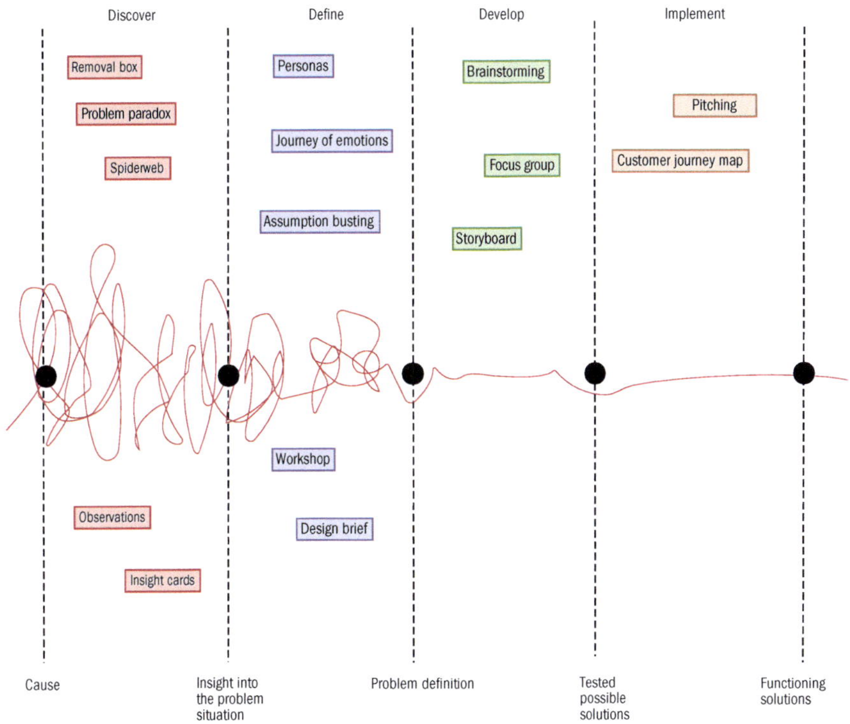

FIGURE 3.9 Make your own roadmap

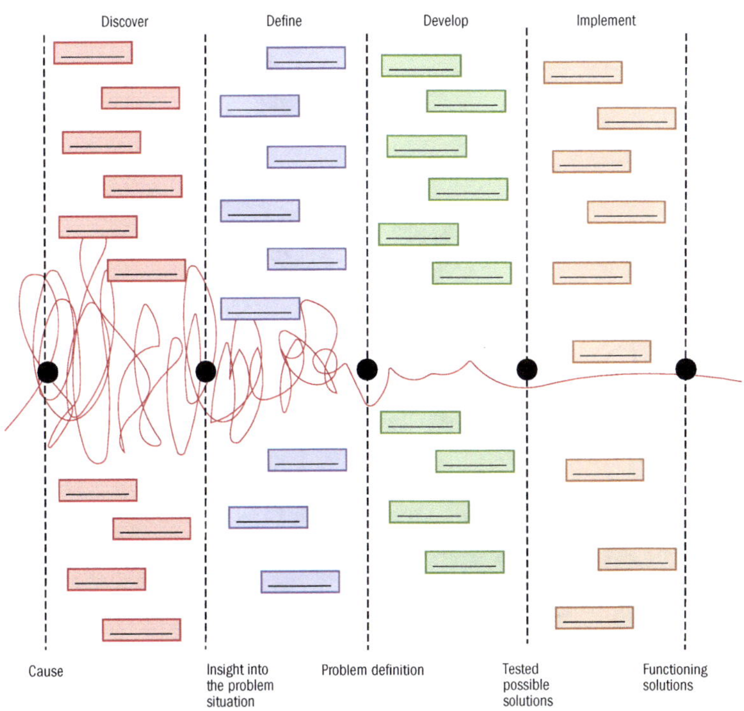

ITSN

ITSN is a medium-sized IT service provider mainly active in the south of The Netherlands. The company went through an intensive project: a complete design process from discovery to implementation. What was special about the design process used at ITSN was that 36 out of 60 employees became actively involved in various design teams. By involving employees, innovation was not limited to the management team or the innovation department.

Each design team was given the name of a racing team. The team that at the end of the process had shown the most progress, were allowed to go to the Formula 1 race in Spa-Francorchamps. Every three weeks there were awards, for example: the most visible team in the organization, the team that involved customers the most, the team with the most unconventional results, the team that showed the most guts and the team that was most talked about at the coffee machine.

Koen van Valderen, project manager at ITSN and team captain of the 'Williams' team, says about this competitive aspect: 'I was skeptical at first. I am not very competitive by nature.' His opinion gradually changed: 'Working in a competitive format was stimulating and was a good way to stay focused on deadlines and strengthened the team spirit within our design team'. And by offering a prize as an extrinsic motivator, the management team showed their

appreciation for the effort made by the teams.

Looking back on the design process, it was ‘great to see how internally oriented colleagues got new insights from focusing on our customers, and how confronting their feedback sometimes was’, says Koen van Valderen.

QUESTION

In this case study, various examples are given for positively stimulating a design team in a design project. Can you come up with examples for stimulating team members to get the best out of themselves and increasing team spirit within an ongoing or completed project?

‘Working in a competitive format in the design team was stimulating. It was a good way to stay focused on deadlines and strengthened the team spirit within our design team.’

— Koen van Valderen, project manager at ITSN.

4
Design thinking is a tool box

This chapter contains a collections of tools that are used to apply design thinking in a practical sense.

This chapter provides answers to questions such as:
- Which tools can I use when applying design thinking in a project?
- What alternative tools are there (with which I have not yet practiced)?
- Which tools suit which fundamental attitude?

The tools that are used in the design process (and therefore in the cycle of design thinking), reflect the six fundamental attitudes that are needed to successfully apply design thinking as a problem solving strategy. The tools are based on the fundamental attitudes and help you to imagine, experiment, cooperate, empathize, work integrally and think flexibly. Thereby contributing to the way of thinking and working used in design thinking. Those who have gone through chapter 3, while working on a project and have used the given roadmap, have already worked with several of these tools. This chapter offers a variety of tools to build your own roadmap and with that, design your own design project.

The tools in this chapter are both extensive guidelines, such as INTERVIEWS or BRAINSTORMING, as well as simple tools such as filling in a SPIDERWEB or making a MOOD BOARD. Not all tools are worked out to the same extent; sometimes we mention only the most important characteristics and refer to an external source. At www.designthinkinginternational.noordhoff.nl you can find the various fill in forms (also in digital format) and some completed forms are included as examples.

To be able to use this chapter as a practical reference, the tools are in alphabetical order. But because we also want to keep inspiring you, we have also included an overview in which the tools are subdivided per fundamental attitude. The majority of the tools can be applied in all phases of the design process and in most cases, they also interface with more than one single fundamental attitude; that is precisely what makes them so interesting for applying them in a design project.

TABLE 4.1 Classification of the tools according to the six fundamental attitudes

Think flexibly
Brainwriting Brainstorming COCD box Decision matrix Post-its Highlighter How Might We Insight cards Problem paradox
Work integrally
Business Model Canvas Design critique Scenarios Spiderweb Stakeholder map Watchtower Removal box
Empathize
10 minute Google Empathy map Focus group Interviews Observations Journey of emotions Shadowing User diaries Questionnaires
Cooperate
Quotation process Pitching Roadshow Design project roadmap Role fulfillment Team meeting Workshop
Imagine
Character profiles Customer journey map Personas Magazine cover Mood board Storyboard
Experiment
Assumption busting Desktop walkthrough Feedback form prototypes One-hour prototype Role play Rules of thumb

10 minute Google

Fundamental attitude: EMPATHIZE

What is it?

A good way to quickly step into the shoes of users or customers, is to take ten minutes to search the internet from *their* perspective. This method of exploring their world is a simple way to start increasing your empathy. After this, you can use other tools. 10 MINUTE GOOGLE is also good to use when developing PERSONAS or when you want to get to know and understand stakeholders or the client.

How does it work?

- Make sure you know whose perspective you are taking and what that perspective is. What is this person doing? What website(s) does he visit and what newspaper(s) does he read? What interests does he have and what questions would he like to see answered in your design project?
- Browse the internet for ten minutes with this perspective. You do not need to actively keep track or write down anything; just ten minutes of getting under the skin of the other person is enough!

NOTES

Assumption busting

Fundamental attitude: EXPERIMENT

What is it?

Assumptions can seem so logical that you do not question them. Right from the start, dare to question what is 'logical or normal', in that way, you come to new insights and discover new and different ways to proceed.
Assumptions include:

- something that seems impossible to do: 'people cannot fly';
- something that works via rules or conditions: 'packages do not fit in the letterbox';
- something that people strongly believe in: 'bugs are a problem'.

With this tool, you question everything about your problem and can test how others think about it. Assumption busting is a simplified form of HOW MIGHT WE.

How does it work?

- Formulate at least twenty assumptions about your design problem that seem completely logical and normal. Let every member of the design team do this individually.
- Question assumptions: What if people can fly? What if all packages fit through the letterbox? What if bugs are the most useful pets you could wish for?

NOTES

Brainstorming

Fundamental attitude: THINK FLEXIBLY

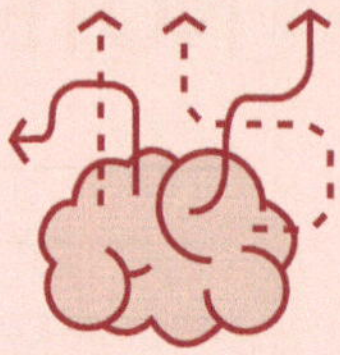

What is it?

Brainstorming originated in the New York advertising world of the 1950s. The idea behind brainstorming is that a quantity of ideas leads to a qualitative, usable (design) product. For brainstorming you need to apply a set of rules in order to come to useful insights or ideas:

- A facilitator guides the brainstorming process. A facilitator defines the scope of the problem for which ideas are conceived and makes sure that all group members have a say in the proceedings. The facilitator ensures that ideas are not dismissed; which is, in fact, deadly for generating nonsensical, illogical or absurd ideas. It is these 'crazy' ideas that provide energy for the group and afterwards often prove to be the seeds for the best ideas.
- All ideas are accepted and noted down (by an assistant).
- Building on other people's ideas is essential to get more out of the group. Because everyone reacts to each other, brainstorming motivates the group members to go one step further in their thinking process (1 + 1 = 3).
- Ideas are not initially ordered or prioritized. It is only after brainstorming for a while, that ideas are prioritized, for example, based on originality or novelty.
- Combining ideas provides usable concepts to keep advancing the design process.

How does it work?

The basic steps for a brainstorming session are:

- Make sure there is sufficient material (whiteboard, post-its, stickers, markers, paper, tape, pins, etcetera) and create the right conditions: turn off all phones, fresh air, plenty to eat and drink.
- Appoint a facilitator for the process and an assistant for writing down, sketching or otherwise portraying all ideas.
- Provide a clear and specific question and split it up to sub-questions.
- Do not criticize ideas from others and yourself. All ideas are accepted and are noted.
- Don't get stuck in your own ideas; 'hitch a ride' on other people's ideas.
- The words 'Yes, but...' are prohibited. Always correct each other by saying 'Yes, and...'. This small change causes an idea to be expanded instead of being dismissed at the outset. Does it seem like the session is chaotic and going in all sorts of different directions? Great, that's fine, it is all about getting many different ideas.
- Set a time limit per brainstorming session (thirty minutes) and per sub-question (maximum five to ten minutes).

Brainstorming may stop after the above-mentioned phase, or things can be worked out in more detail:

- Cluster the suggested ideas into similar ideas or concepts.
- Give participants a number of stickers to stick on the ideas they want to see worked out (dot-voting).
- Continue to brainstorm on the ideas with the most stickers.
- End the session by having the facilitator provide a summary of the process and the ideas that need to be worked out.
- Agree on feedback to be given about what happened to the ideas afterwards.

Pitfall 1: A brainstorming session can get stuck. The facilitator will then need to use 'stimulants'. This can be achieved by literally moving people or for example, using randomization. The idea behind randomization is that creativity is encouraged when people are stimulated via completely arbitrary cases, for example:

- Enter a random idea in the brainstorming session. Open a book and pick out a word (blindly). You can also randomly grab an object.
- Now, associate the word or object with the subject of the brainstorming session.

Pitfall 2: For some people, a brainstorming session can possibly be quite overwhelming. To get ideas from introverted people onto the table, you can start the brainstorming session individually:

- Have everyone write down their ideas on cards for themselves.
- Make an inventory of the ideas by having people present their cards
- During the explanation of the ideas, others can supplement their own list as they come up with new ideas.
- The organization and prioritization of the submitted ideas are done by means of individual ranking (so everyone makes their preferred list), and finally by vote.
- See also the regular brainstorming process.

Yes, but...

Do you want to experience how counterproductive 'yes, but...' can be? Do the exercise on page 27.

Brainwriting

Fundamental attitude: THINK FLEXIBLY

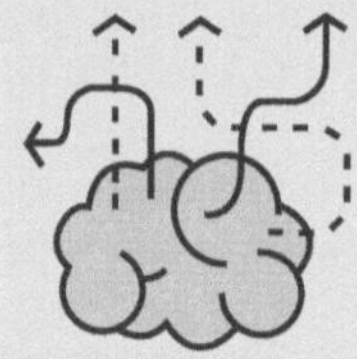

What is it?

Brainwriting is the silent variant of brainstorming. Instead of 'calling out' ideas, they are written down. When writing down ideas everyone has a chance to offer ideas and they are formulated more thoughtfully. Brainwriting aims for the same cumulative effect as brainstorming (ideas are enhanced by reactions to reactions), but in this case in silence.

How does it work?

- Define what the brainwriting is all about.
- Give all participants in the brainwriting session a copy of the fill in form.
- Have them write down the subject of the brainwriting session at the top.
- Everyone writes down three ideas in the boxes for 'round 1' in two minutes.
- Pass the forms on (to the person left of you, or think of another creative way to do that).
- Everyone reads the ideas of the others from round 1 and in response, writes down three new ideas in the boxes for 'round 2'. Allow three minutes for this.
- The forms are then passed on until all boxes are full. As a facilitator, emphasize that there must always be a new response to the previous round.
- When the form is fully completed, pass it one more time. Now, everyone chooses the three ideas that appeal the most and encircles them.
- Cluster similar ideas (and possibly reformulate them into concepts). See POST-ITS for this.
- Share the three ideas (or concepts) that will be elaborated with the group.

FILL IN FORM 4.1 Brainwriting

Round	Subject:		
1			
2			
3			
4			
5			

Business Model Canvas

Fundamental attitude: WORK INTEGRALLY

What is it?

The Business Model Canvas was developed by Alexander Ostenwalder. It is a commonly used way of gathering and summarizing information about the desirability, feasibility and viability of a possible solution.

The value proposition is central to the Business Model Canvas: this is the value that the solution creates for the customer, in the form of a product or service. The right side of the canvas is about the customers for whom the value is to be created; the left side is about how to get the value created. At the bottom of the canvas, there is room for recording the costs for creating the value proposition and the revenue that the solution will need to generate.

How does it work?

- Start by entering the value proposition: What is the value that the solution, in the form of a product or service, creates for the customer and what is the distinguishing power of this solution (compared to the competition)?
- Then complete the form in *random* order:
  - Customer segments. Who are the users of the value proposition?
  - Customer relationships. In what way is the relationship with the users actively maintained?
  - Channels. Through which channels are users contacted?
  - Key resources. Which resources can be used for the realization of the value proposition?
  - Key activities. Which activities are important for the realization of the value proposition?
  - Key partners. Which partners are of strategic importance for the realization of the value proposition?
  - Revenue streams. What sources of income will be generated by the value proposition?
  - Cost structure. Which costs are involved in creating the value proposition?

Want to quickly check the viability of a business model?
Ostenwalder (2010) devised seven questions to determine whether a business model is viable:

1 How much do switching costs prevent your customers from churning?
2 How scalable is your business model?
3 Does your business model produce recurring revenues?
4 Do you earn before you spend?
5 Are there others who can do (part of) the work?
6 Does your business model provide built-in protection from competition?
7 Is your business model based on a game changing cost structure?

Read more about the Business Model Canvas?

Read Alexander Ostenwalder's book *Business Model Generation* (2010)

FILL IN FORM 4.2 Business Model Canvas

Key partners:

Key activities:

Key resources:

Value propostition:

Customer relationships:

Channels:

Customer segments:

Cost structure:

Revenue streams

Bron: strategyzer.uservoice.com

NOTES

Character profiles

Fundamental attitude: IMAGINE

4

What is it?

A character profile helps to zoom in on the target audience that the design process focuses on (and which target group you are *not* targeting). Know who your customers or future users are, what is going on with them, what wishes and needs they have and what communication channels they use to express themselves.

How does it work?

- Identify the different target groups. Take the following into consideration:
  - demography, socio-economic factors, preference for products or services;
  - lifestyle, challenges, norms and values;
  - influencers (or heroes), internet behavior and media behavior (social media, newspapers, TV programs).
- Do not only make character profiles for average customers, but also for extreme users: they can expose important aspects of the problem.
- Give the different user groups a catchy name and a face.
- Make a visual overview of the different character profiles.
- If you have to make decisions about ideas, concepts or prototypes, check out the different character profile perspectives: How would a frustrated user respond? What would an early adapter (someone who loves innovations) notice?
- Do you want to describe the defined character profiles in more detail? Check out how PERSONAS are worked out.

NOTES

4

COCD box

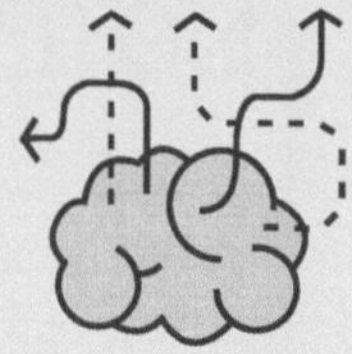

Fundamental attitude: THINK FLEXIBLY

What is it?

The COCD, Center for the Development of Creative Thinking in Belgium, developed a handy tool to prioritize ideas or solutions based on feasibility and originality. The COCD box distinguishes three categories of solutions:

1 Blue solutions are feasible 'normal' solutions of which examples are usually already available ('NOW'). Blue solutions are easy to implement, take relatively little effort to make, have few risks and enjoy wide support in the organization.
2 Red solutions are feasible original solutions ('WOW'). They are innovative, break through existing patterns, give energy and are distinguished from other solutions. These are ideally, the ideas or solutions which can be developed via the design process.
3 Yellow solutions are original solutions that are not (yet) feasible ('HOW'). There is a challenge in these ideas; they mainly show a vision. They are dream images, possible solutions for the future.

How does it work?

- List all the conceived solutions or ideas.
- Give each team member five red stickers, five blue stickers, and five yellow stickers.
- Team members give solutions one or more stickers. NOW solutions get blue stickers, WOW solutions get red stickers and HOW solutions get yellow stickers.
- The solution with the highest quantity of red stickers gets the highest priority for the design team.

FIGURE 4.1 COCD box

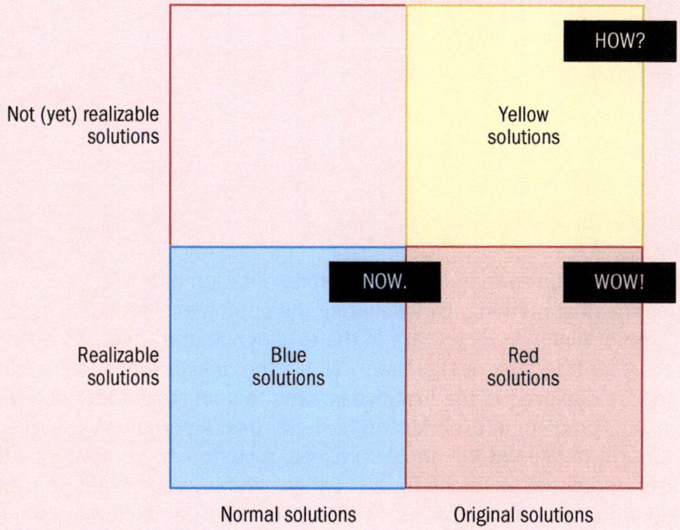

Source: Centrum voor de Ontwikkeling van het Creatief Denken

Customer journey map

Fundamental attitude: IMAGINE

What is it?

A customer journey map is a visual representation of how a user experiences something. By visualizing the customer's experience, you find out where 'highs and lows' are in the experience and what has an influence on the client and the design team. Within the design process, the customer journey is deployed in the first two phases to gain insight into how a particular problem is experienced from the user's perspective. In the third and fourth phases of the design process, a customer journey helps to imagine potential solutions for a client or can be used to test possible solutions among stakeholders. You can also use the customer journey to compare the existing situation with the desired situation and determine where improvements are possible.

How does it work?

- Determine the phases that the user goes through, before, during and after using the product or service or experiencing the problem.
- Investigate which different activities the user performs per phase and then complete the form. Set the time of purchase, use of the service or the moment the problem is experienced, somewhere in the middle.
- Trace what the user thinks and feels during each phase and activity.
- Consider whether and through which channels, the user is in contact with the organization during the different phases.
- Provide the most important insights from the user's perspective per phase, in 'highs and lows'.
- You already have ideas for possible solutions while you are in the discovery or definition phases? Don't ignore them, but write them down for later.

FILL IN FORM 4.9 Customer journey map

Phase 1: ______	Phase 2: ______	Phase 3: ______	Phase 4: ______	Phase 5: ______	Phase 6: ______
Activities (What does the customer do?):	Activities (What does the customer do?):	Activities (What does the customer do?):	Activities (What does the customer do?):	Activities (What does the customer do?):	Activities (What does the customer do?):
Thoughts & emotions (What does the customer feel and think?):	Thoughts & emotions (What does the customer feel and think?):	Thoughts & emotions (What does the customer feel and think?):	Thoughts & emotions (What does the customer feel and think?):	Thoughts & emotions (What does the customer feel and think?):	Thoughts & emotions (What does the customer feel and think?):
Touchpoints:	Touchpoints:	Touchpoints:	Touchpoints:	Touchpoints:	Touchpoints:

The six most important highlights:

1 ______ 4 ______
2 ______ 5 ______
3 ______ 6 ______

The six most important lows:

1 ______ 4 ______
2 ______ 5 ______
3 ______ 6 ______

Solutions for the next phase:

______

Decision matrix

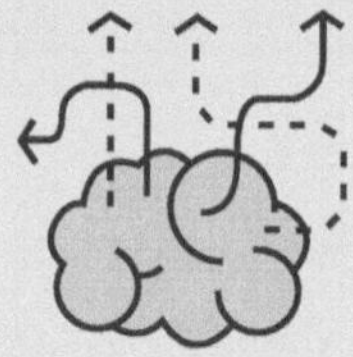

Fundamental attitude: THINK FLEXIBLY

What is it?

A decision matrix helps you to prioritize ideas or concepts on the basis of an actual comparison, rather than based on feeling or intuition. A decision matrix is a table in which an idea or concept must comply with certain criteria, and is converted into a yardstick with weighting factors. An idea or concept is assessed based on these criteria/weighting factors; the idea or concept with the highest score would be developed further.

How does it work?

- Make a list of all criteria that you want to take into account in determining your choice. Examples of criteria are: price, quality, sustainability, fashion sensitivity, comfort, availability, exchange options and upgrade options.
- Make a scale of 1 to 5 per criterion. For example, for price: 1 = expensive, to 5 = cheap; And for comfort: 1 = not nice to use or not comfortable, to 5 = very nice to use or very comfortable
- Determine the weighting factor for each criterion and express it in a number. A weighting factor is the relative importance of a criterion. As the price of goods or services is very important, it can be two or three times the weighting factor of, for example, sustainability.
- Make a matrix with the options from which you want to make a choice on the Y-axis and on the X-axis the factors on which you base the different options you want to score.
- Score each option in the matrix from 0 to 5 and multiply the score with the weighting factor.
- Add up the scores per option to arrive at a final choice.

FILL IN FORM 4.3 Decision matrix

Decision matrix for: ____________________

	Criterion: ______ Weighting factor: ______	Critera: ______ Factor: ______	Critera: ______ Factor: ______	Totals
Option 1: ______				
Option 2: ______				
Option 3: ______				
Option 4: ______				
Option 5: ______				

Design brief

Fundamental attitude: THINK FLEXIBLY

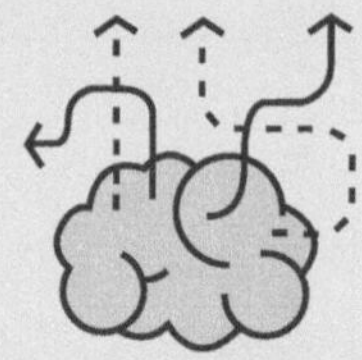

4

What is it?

A design brief marks the end of the first two phases and provides a launch for the last two phases of the design process. The design brief must stimulate the team to continue with the design process in a productive manner. It presents the core design challenge. A design brief is an organized display of the information collected from the first two phases of the design process. It includes a concrete description of the design problem. It describes cause-effect relationships that have been discovered and displays the research results and figures that substantiate the problem.

Design thinking takes the organizational perspective into account but looks at the problem from the customer perspective. The design brief, therefore, translates the insights from the first two phases of the design process to the user perspective, thus also focusing on the user in the last two remaining phases. When preparing the design brief, don't formulate the answers for the problem, however tempting that may be.

How does it work?

- Complete the fill in form.
- Check the completed form on the following points:
  - Has the problem been described clearly enough, without too much direction for a possible solution?
  - Are the formulations positive and constructive?
  - Have the objectives been made specific?

FILL IN FORM 4.4 Design brief

Project name: ______________________

Date: ______________________

Design problem or challenge in max. 200 words:

Goals:

What impact should the solution have on the problem?

About the organization:

Timing/planning and possible barriers for the design team:

What needs to be organized to facilitate the design process?

Who needs to be involved?

Who needs to be reported to?

Pitfalls or critical issues to take into account:

Other usefull information:

Design critique

Fundamental attitude: WORK INTEGRALLY

What is it?

A design critique can be seen as a sounding board of people who are motivated to find a solution for the design problem but who can also look critically at the existing insights, the initial ideas or possible solutions and dare to communicate honestly about it. A fresh look ensures that weaknesses, for which the design team may have a blind spot, can come to the surface. The design team must be prepared to take one step back from the process if criticism from the meeting requires this. The thought behind a design critique is that if, at an early stage, you ask others to look critically at ideas/possible solutions and ask them to explore the limits of the design problem, there's a greater chance that the final solution will be successful. A design critique ensures fine-tuning of the design process. It is often deployed to get the design process moving on to the next phase.

How does it work?

- Provide a facilitator with the right tone of voice and a positive attitude. Ideally, this is someone who is not too attached to the insights, ideas or solutions and is able to translate negative feedback into positive, constructive feedback that helps the design process to move forward.
- Organize a session with four to six people who are (possibly) critical towards the design process. Do not think in terms of functions or hierarchical positions of people; determine who can make the best contribution.
- Create an atmosphere where everyone feels free to express and share his or her opinion. This means that a safe, inviting atmosphere must be created.
- Formulate a number of specific questions that you want to see answered in the session.
- Make ideas about the problem tangible, print out sketches of possible solutions or work with four to six prototypes: make it as visual as possible.
- Hang the various prototypes/sketches on the wall or lay them out on a table. If possible, give a demonstration of the prototypes.
- Start an open discussion and let the conversation run its course. Ensure that you receive an answer to the pre-formulated questions at the end of the meeting.
- Indicate how criticism will be dealt with or processed and how and when the received criticism will be communicated back to the participants.
- Schedule a follow-up appointment if necessary.

Peer review as an alternative or in addition to design critique
Organizing a peer review is a good addition or an alternative to, a design critique with stakeholders. A peer review is similar to the design critique, but among peers, instead of inviting directly or indirectly involved persons to give feedback. The design team reviews, together with colleagues or fellow subject matter experts, on the current status of the design process as well as on insights and ideas that have been found.

Design project roadmap

Fundamental attitude: COOPERATE

What is it?

A roadmap provides a picture of what the design project will look like, which tools are going to be used and provides a timeline estimate. A roadmap provides guidance for the design team and the stakeholders. Compare the use of a roadmap with using an actual road map in which you map out a route from A to B: you know along the way you can make detours or stop earlier or later, but you still have a global overview of the route.

In order to be able to draw up a roadmap, you go through the four phases of the design process very quickly. Among the questions to be asked are:

- What is the lead time of the design project?
- Which steps can be distinguished within the phases?
- Which milestones do we establish?
- Which tools are used in which phase?
- How should the design team be put together?

After each milestone, you reflect on the chosen path and steer the design process based on advancing insights. A roadmap is therefore not definitive, but is constantly being fine-tuned as the project progresses. Do you have to convince a client of the roadmap that you have drawn up in the form of a quote? See also: the QUOTATION PROCESS.

How does it work?

1 **Make a quick scan of the cause for the project.** Planning a design process starts with performing a quick scan. This is a mini version of the discovery phase. Take one or two hours for this and collect information about the cause for the design process and any other issues that you consider important:
   - Go through the cycle of design thinking a number of times in a very short time until you have a basic understanding of the reason behind the design process.
   - Use the information for initial planning of the design process.

2 **Imagine an initial roadmap**. A roadmap is a representation of the best route to follow, given the current knowledge and the estimated lead time:
   - Make an initial roadmap outlining the tools you want to use, planned per phase and per step, of the design process.
   - Establish milestones that you want to achieve. The milestones are the markers for the next phase of the design process.

3 **Form or evaluate the design team**. After the design process is mapped out in general terms, the design team needs to be set up:
- Use the ROLE COMPLETION tool to form a design team.
If there already is a design team, assign the roles and take a critical look if you have an optimally functioning team. Plan a session with the team to get and discuss a first version of the roadmap and complete the roadmap together.
- Divide tasks, finalize milestones and plan them in time.

4 **Plan stakeholder involvement:**
- Consider where others are involved in the project:
  - At what times and to what extent do you need management or do you want to inform the client?
  - When do you expect input from colleagues in department X, Y and Z?
  - Which users will you involve and how?
  - Which fixed communication moments will you need to plan? Is this a physical meeting or is this via a medium (newsletter, email, social media)?

5 **Schedule moments for advancing insights**. You cannot predict how and what, but a roadmap will always change because of advancing insights:
- Plan for evaluations of all the completed steps during each phase of the design process.

6 **Plan for information flows:**
- Choose an information storage system so that during the design project, information is easily available for those involved.

7 **Provide insight into project costs:**
- Determine in advance what the budget is for completing the project. Think of:
  - costs for deploying own staff (employer's costs);
  - costs for deploying third parties (professionals that you hire and their hourly rate, fees, gifts for customers or users participating in testing, etcetera);
  - costs for deploying tools (prototypes, materials, systems, external space);
  - costs for fun (inspiration session with the project team, final dinner or party).

FILL IN FORM 4.14 Roadmap for the design process

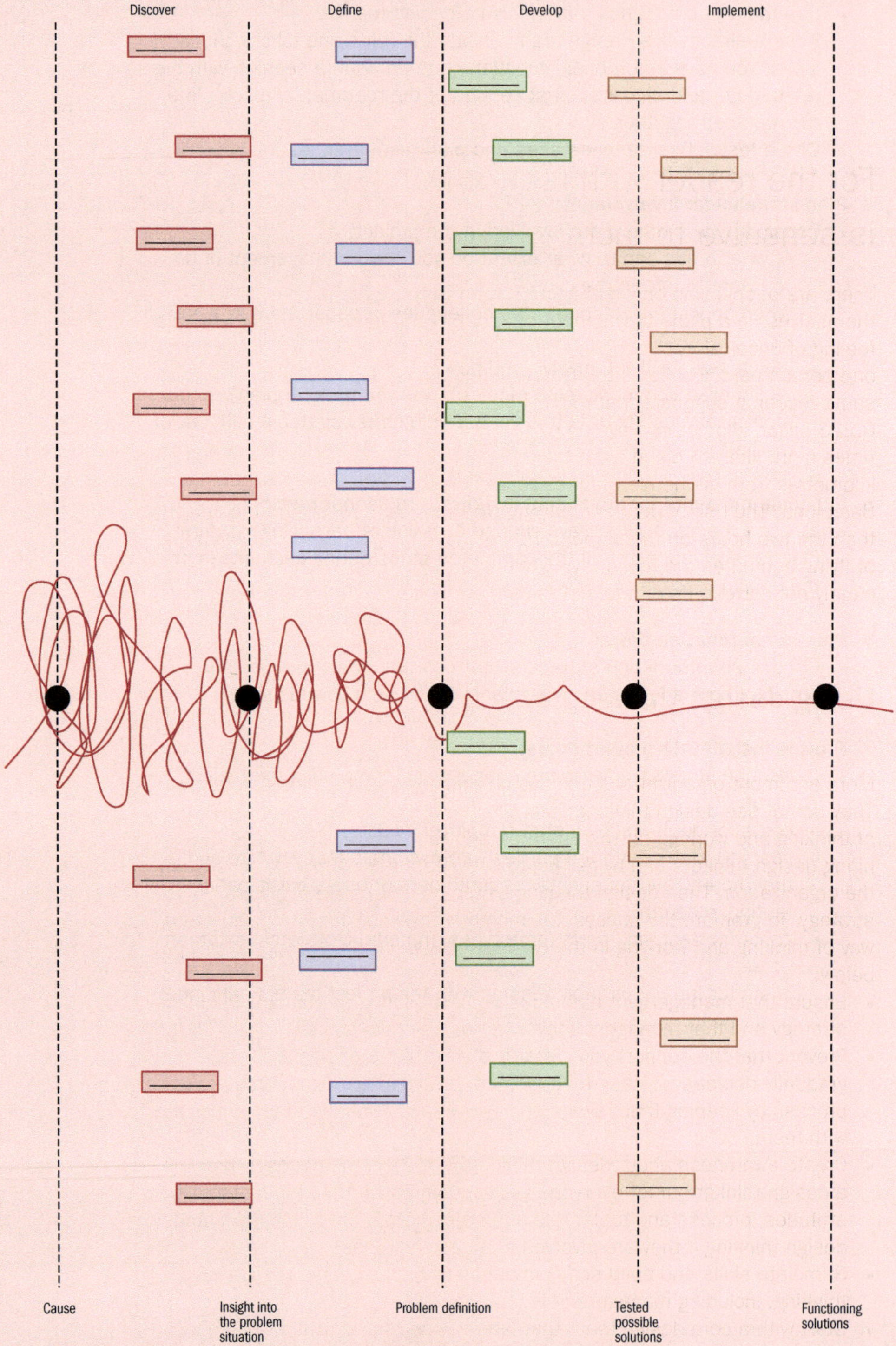

4

NOTES

4

Desktop walkthrough

Fundamental attitude: EXPERIMENT

What is it?

A desktop walkthrough is a way of experiencing a product or service from the user's perspective. During a desktop walkthrough, a physical experience is played out in the smallest detail in order to gain insights into the key issues of implementing the product or service, based on which improvements can be made.

The desktop walkthrough consists of:

1 Determining the script
2 Creating an environment
3 Adding complexity
4 Drawing conclusions

How does it work?

1 First determine the script for the actual experience that you want to replay:
   - First describe the problem you want to use for the script. Example: the problem being worked on is 'there are too many complaints'.
   - Decide which script you will work out for the problem you want to solve or the solution you want to test. Example: a script that you want to elaborate on is: 'complaint handling after submitting a complaint form'.
   - Work out the script in a STORYBOARD, a CUSTOMER JOURNEY MAP or a JOURNEY OF EMOTIONS.
   - Choose figures that represent the customers (experiencing the problem) and divide the roles among the members of the design team. LEGO figures are ideal for this, but you can of course also use your own. Examples: the dissatisfied customer, her girlfriend who came along, a random passer-by, an employee who has 'caused' the complaint, an employee who records the complaint, other employees, etcetera.

2 Creating an environment:
   - Scale model the environment where the script takes place. You can use LEGO for this or other material.
   - Play the script. There may be matters that are not correct in the script. Adjust these and play the script again until correct.

3 Add complexity:
   - During the repeated replay of the script, things come to mind that have not been thought of before.
     Example of 'reducing complaints': What if all the complaint forms were used up? What happens if we add more passers-by?

- Also, try to get more abstract things on the table.
  Example of reducing complaints: How long must the customer with a complaint wait for a response? What happens if the seller resolves the complaint himself immediately?

4 Draw conclusions:
- Draw conclusions after replaying the experience:
  - What are the 'mistakes' in the experience?
  - Have we come to new insights?
  - What new ideas have we received?
- Provide feedback to stakeholders by adapting the DESIGN BRIEF.

Empathy map

Fundamental attitude: EMPATHIZE

What is it?

It can be difficult to put yourself in the position of a future user. An empathy map helps to investigate what the user thinks, does and feels, both within the context of the problem and where it occurs as well as outside the user's own environment.

How does it work?

- Fill in the form. It helps you to visualize a typical day of a typical user.
- What does this person teach you? What surprises you? What consequences does this have for your design process or for your further research?
- In preparation, you may have already done 10 MINUTE GOOGLE. See also: PERSONAS and CUSTOMER JOURNEY MAP.

FILL IN FORM 4.5 Empathy map

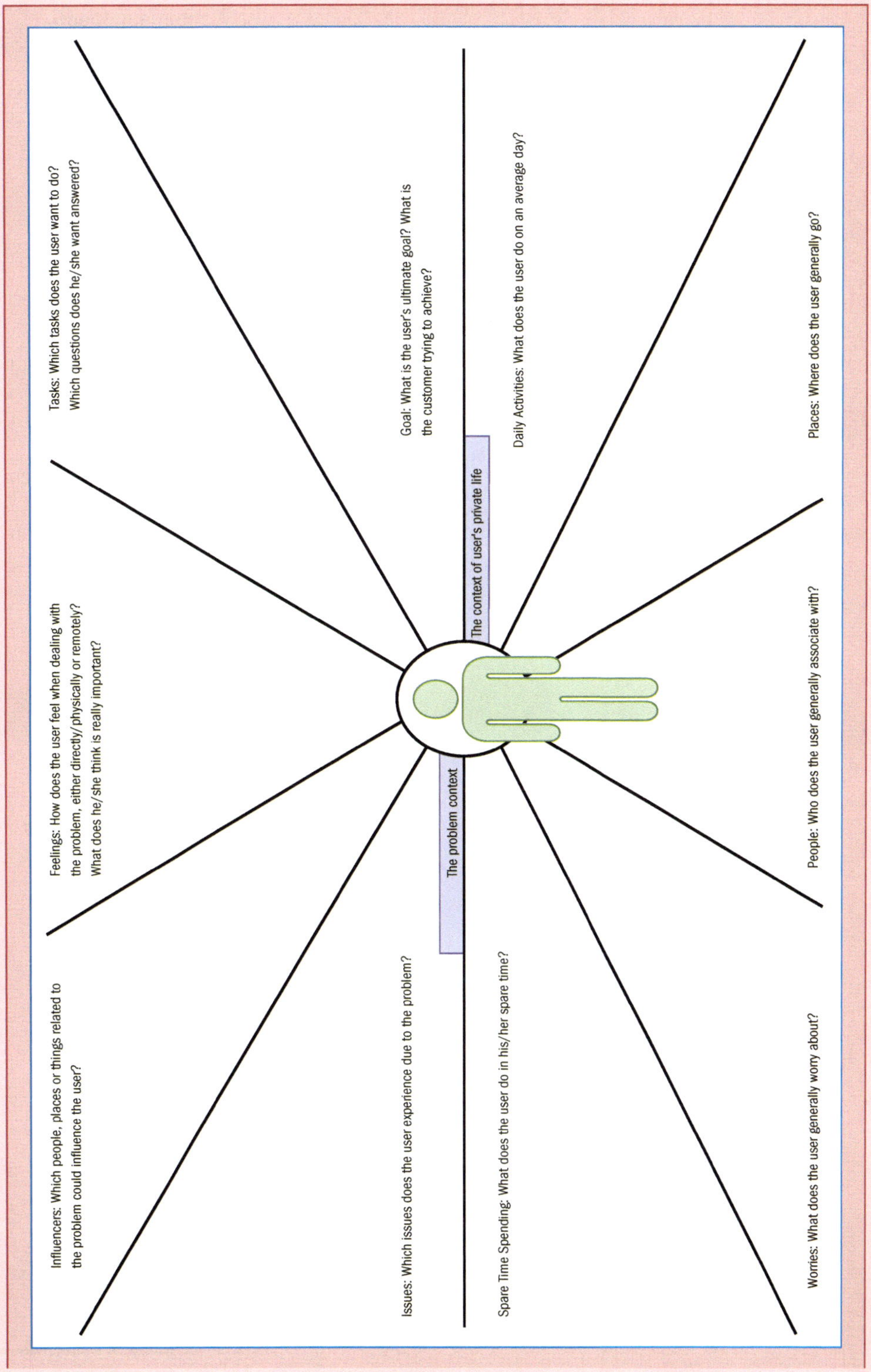

Feedback form prototypes

Fundamental attitude: EXPERIMENT

What is it?

You would want to test prototypes with users (or others) during the design process. You can use the feedback form to record which elements of the prototype you want to keep, which you want to get rid of, which you want to develop (i.e. strengthen or moderate).

How does it work?

- Complete the fill in form for each prototype.

FILL IN FORM 4.6 Feedback form prototypes

Prototype name: ______________________________

What to keep:

____

What can be omitted/discarded:

____

What needs to be developed/strengthened:

____

What to attenuate/weaken:

____

4

Focus group

Fundamental attitude: EMPATHIZE

What is it?

A focus group is used to discuss a variety of topics as seen from the perception of a specific group of people. A focus discussion provides a lot of concrete information about opinions, attitudes, ways of thinking and working, needs, wishes and priorities, in a short period of time. Focus groups are used in the definition phase and in the development phase to get feedback on ideas, concepts or prototypes. But you can also use a focus group in the implementation phase.

A focus group consists of five to ten people, a facilitator and an observer. The facilitator sets up the agenda, makes the right preparations, asks questions during the session and keeps asking questions. The observer takes notes, observes any underlying messages, helps the facilitator with group dynamics and evaluates together with the facilitator afterward. Note: a focus group is not the same as a group interview. It's not about the interaction between the interviewer and the interviewee, but more about group dynamics and the discussions that arise among the participants.

How does it work?

- Determine the agenda based on the topic and on the phase of the design process for which you will be using the focus group.
- Determine how many people you need and invite them. Also determine if your target group is intrinsically motivated (colleagues, a platform with like-minded people, etcetera) or that you assume extrinsic motivation and provide a (minor) compensation for their input.
- At the beginning of the session, ensure that there is a group dynamic that is safe, creative and open. Go through the basic rules (confidentiality, transparency, listening to each other, positively reinforcing each other).
- Provide a suitable space that has the right atmosphere and technical facilities (for example for recording or playback). Make a room arrangement (for example, a circle with or without tables) that on the one hand, radiates safety and coziness and on the other hand, enables the facilitator to note everyone's contribution.
- Let the group dynamics do the work. A facilitator who takes the lead too much, risks looking for confirmation of the assumptions about the problem or positively reinforcing the already devised solutions. Ensure there is an open structure (see also INTERVIEWS) and do not take too many notes during the discussion (make video or audio recordings for subsequent analysis). As the facilitator, ask open questions: Can you tell

me something about the last time that you...? What was nice about it? What was frustrating about it? How could we improve...? What would you do differently?

- Conclude the session with (possible) follow-up appointments and thank everyone for their time.
- The facilitator and observer evaluate the session as quickly as possible and discuss the results:
  - Have all of the questions been answered?
  - What were the most important insights?
  - Did new things emerge that previously did not get any attention?
  - Who should know about the new insights? Does this have any consequences for the DESIGN BRIEF and/or the ROADMAP?

Unfocus group

What is it?

An unfocus group does not focus on an idea, concept or prototype, but is all about getting to know a group of customers better as people, regardless of the problem you are trying to solve in the design process. With an unfocus group (with customers or users, for example) you can feed a random group discussion, without an agenda and without manuals or advance preparation. If you have been engaged in the design challenge for a longer period of time as a design thinker, there can certainly be issues that you no longer see clearly and blind spots may arise. An unfocus group can then provide surprising insights into unfulfilled expectations and wishes that you can respond to in the design process.

How does it work?

When deploying an unfocus group, make sure that you:

- Resist the temptation to provide some structure or ask leading questions about the problem;
- Work anonymously if possible. Participants will be even more honest in sharing insights that way.

Highlighter

Fundamental attitude: THINK FLEXIBLY

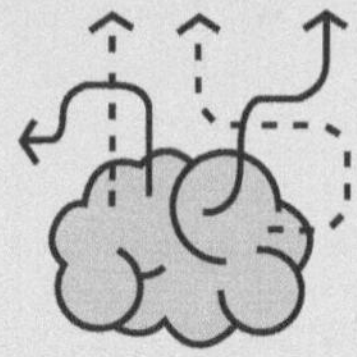

What is it?

You can accentuate the most important insights from a large number of documents, by highlighting these important insights using different color markers. Use those different colors to distinguish insights that are easy or difficult to tackle or about which you have doubts. Highlighting is relatively fast and helps prioritize insights according to importance.

How does it work?

- Print the document with the various insights.
- Highlight the most important insights with different color markers:
  - green: insights that are easy to tackle;
  - orange: insights that are difficult to tackle;
  - yellow: insights that you still have doubts about.
- Determine in the design team which insights will be used in the design process.

NOTES

How Might We

Fundamental attitude: THINK FLEXIBLY

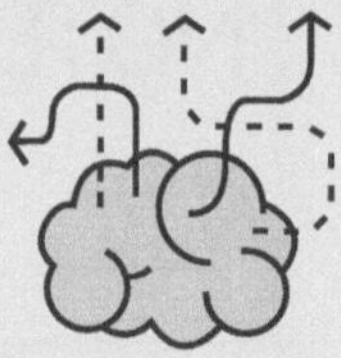

4

What is it?

In design thinking, the principle of How Might We, is usually applied in the transition from the definition phase to the development phase. You can recognize several fundamental attitudes:

- How stands for: There is always a solution to a problem, we only have to find it.
- Might means 'could': let's experiment with it.
- We means that you cannot solve a problem on your own, but you will have to collaborate with others.

With How Might We you translate the obtained insights from the discovery phase into different action-oriented questions, each with a different perspective.

Ask How Might We questions in which you:

1 identify unexpected possibilities;
2 emphasize what is positive about the insight;
3 make a comparison with something else;
4 split the insight into several elements;
5 omit the negative aspect from the insight;
6 have doubts about assumptions concerning the insight.

How does it work?

- Use the different perspectives to formulate How Might We questions.
- Use the format provided to record them.

FILL IN FORM 4.7 How might we

Insight: ______________________________

How can we identify the unexpected?

____
____
____
____

How can we emphasize the positive aspects?

____
____
____
____

How can we compare it to something else?

____
____
____
____

How can we break it down into parts?

____
____
____
____

How can we do away with the negative aspect(s)?

____
____
____
____

How can we bust the assumptions?

____
____
____
____

Source: leanservicecreation.com

Insight cards

Fundamental attitude: THINK FLEXIBLY

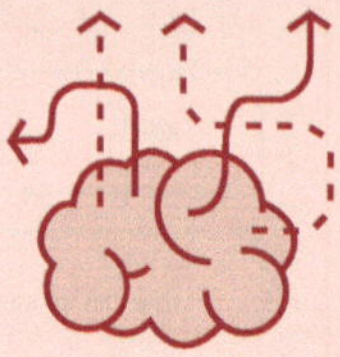

4

What is it?

At the end of the discovery phase, you summarize your findings by formulating various insights. Use the insight cards to make the most important findings concrete.

How does it work?

- Fill in as many insight cards needed until you have all the insights from the discovery phase on paper.
- Complete the forms with sketches or other visualizations.

FILL IN FORM 4.8 Insight card

Project name:

Date:

Insight:

Important additional info (from the past):

Most important people involved internal:

Most important people involved external:

Important information sources for this insight:

Important aspects for the definition phase:

Interviews

Fundamental attitude: EMPATHIZE

What is it?

An interview is an essential tool in all phases of the design process for collecting information. Interviewing future users, clients or other stakeholders is a good way to empathize with them. Observation can be the basis for an interview. Andrew Travers mentions five steps in Interviewing For Research, A Pocket Guide:

1 recruiting
2 preparing
3 conducting
4 documenting
5 synthesizing

How does it work?

1 Recruiting

Carefully recruit the people you need for your research. Who is interesting to interview? How do you find those people and how can you approach them? Make a distinction between insiders that are professionally interesting for the design process and insiders that belong to the political arena of the organization. Involve known and unknown outsiders. When recruiting, pay attention to the following:

- Choose professionals who know the ins and outs of the organization and can provide you with detailed information that is needed for a successful design result.
- Invite people at strategic and tactical levels who are able to create support, goodwill and be fruitful for the design project.
- Do not only choose outsiders who are already familiar with the organization and satisfied with the 'old' product or service. Start looking for future customers, who can take a fresh look at the product or service and the organization in general.
- Make a profile of the people you want to talk to. If you are a recruiter this is certainly essential. For example, profiling can be done based on PERSONAS and CHARACTER PROFILES.

2 Preparing

Which interview setup you choose depends on where you are in the design process: will you use the interview to test a frame, a business (model), a rough or developed detailed prototype? Prepare yourself by making an appropriate interview schedule:

- An interview schedule consists of the main topics that are to be addressed. Do not formulate ready-made questions to avoid forcibly leading the conversation.

- Use models, prototypes and drawings to represent any ideas or hypotheses.
- Prepare the interviewees for what is to come: contact them in advance, introduce yourself and email any background information. Reassure the interviewees that no special preparation is needed.
- Report how the feedback will be used in the design process.
  Finally, also include practicalities, such as defining the location or setting, taking into account noise and loss of non-verbal communication when using technology such as phones or Skype and plan for sufficient travel time.

3 Conducting

Realize that someone is freeing up valuable time for sharing their thoughts, needs, opinions and experiences with you. So make sure you have a good conversation, for yourself and for the interviewee. A good conversation is not the same as a nice and pleasant conversation: ask critical questions, but be neutral; be alert, curious and 'the naive outsider'. That's how you make sure that the interviewee can be the expert who provides you with information.

A number of tips are:

- Use all your communication skills.
  - Use listening, summarizing and questioning as the basis for every interview. Practice your communication skills in advance on your design team or roommates (and note how much more information you collect).
  - Give the interviewee space and do not be afraid of moments of silence during the interview. Creating silence is a not-to-be-underestimated way to get the interviewee to process and perhaps provide additional information. A tactical sip of water or taking some notes is enough to create a moment of peace (meanwhile you can think if the conversation is going in the right direction).
  - The question 'Can you tell more about it?' Always works well to encourage the interviewee and give him or her the feeling that what they are saying is important to you.
  - Do you want to use IMAGINE Then do not use words but images or photos.
- Do not entice the interviewee to come up with solutions for the issue; the interview is not intended for this.
- Do not ask questions about hypothetical behavior: 'How would you travel to Paris?' In practice, people behave differently than they say (or think). So focus on observable behavior: 'How did you travel the last time you went to Paris? '
- Only distill information that is useful for the design process by understanding the observed behavior. Therefore, continue asking questions until you uncover the cause-effect relationship in the behavior. Ask 'why' at least five times.
- If there is a prototype: observe the reaction to this future design and see how the prototype is actually used. It is tempting to ask suggestive and closed questions to confirm your enthusiasm for the prototype. Prevent this by getting new information from this unique person by asking questions that can bust assumptions about the problem, the selected frame, the idea or the prototype.

4 Documenting

- Before interviewing, consider how you want the feedback from the interviews to be documented so that it later provides the information that helps the design process.
- Also consider in advance which method of documentation is most suitable to inform, motivate or convince the other members of the design team who were not involved in the interviews.
- Be careful when summarizing interviews. This can make the conversations seem too 'flat', leaving out details or 'irrelevant' comments which later may be the key to the right design.
- Record conversations: you do not know what you can get out of the interviews later.
- Write down your own thoughts during the interview in the margins and after the interview, take plenty of time to let the conversation sink in: What were the special moments, what insights have you gained from this unique conversation or what non-verbal hints did you get? What do you want to remember from this unique conversation and take back to the design process?
- Use this post-interview moment for self-reflection on your role as an interviewer: Which questions generated energy for the interviewee and what kind of questions did you ask? What do you want to take with you or explore in the next interview?

5 Synthesizing

After conducting the interviews, your head is full of ideas, (perhaps conflicting) information and different perspectives. How do you summarize a pile of different interviews that can inspire or influence insights for the design process? How can you get the feedback from the interviews and the information in your head translated and transferred to the minds of the whole design team? To do this, proceed as follows:

- Remember that with design thinking a conclusion about a design solution does not come from a generalization of all interviews, but arises from the uniqueness of a conversation, a highlight, a moment. Details are lost when analyzing data. It's the authenticity of that one conversation, its carelessness, chaos or the doubts and inconsistencies on the part of the interviewee, which makes the interview interesting.
  So keep far away from general conclusions or recommendations.
- Do not wait until all the interviews have been completed. Keep the design team involved and make sure the rest of the team can already get started based on the results.
- Use the fundamental attitude IMAGINE: tell the story of the interview to your design team and let them respond, ask questions and challenge you. STORYTELLING helps you to get to the heart of the message from the interview.
- Use the interviews to create CHARACTER PROFILES and PERSONAS or fine-tune the already created character profiles and personas.

NOTES

4

Journey of emotions

Fundamental attitude: EMPATHIZE

4

What is it?

If it is not possible or practical to make an extensive CUSTOMER JOURNEY MAP, you can choose making a journey of emotions. A journey of emotions can also be started in preparation for a customer journey map. A journey of emotions is a quick way to visualize the emotions of the user so you can empathize.

How does it work?

- Determine, on the basis of research, what a typical user is experiencing and divide this into a maximum of 16 consecutive steps. Every step represents one activity or action.
- If you have a STORYBOARD, use it.
- Write the sixteen steps on the X-axis.
- Indicate with a dot on the Y-axis how the user feels during each of the sixteen steps. Determine this by using OBSERVATIONS or USER DIARIES.
- Draw a line between the different points and see how the users' journey of emotions has transpired.

FILL IN FORM 4.13 Journey of emotions

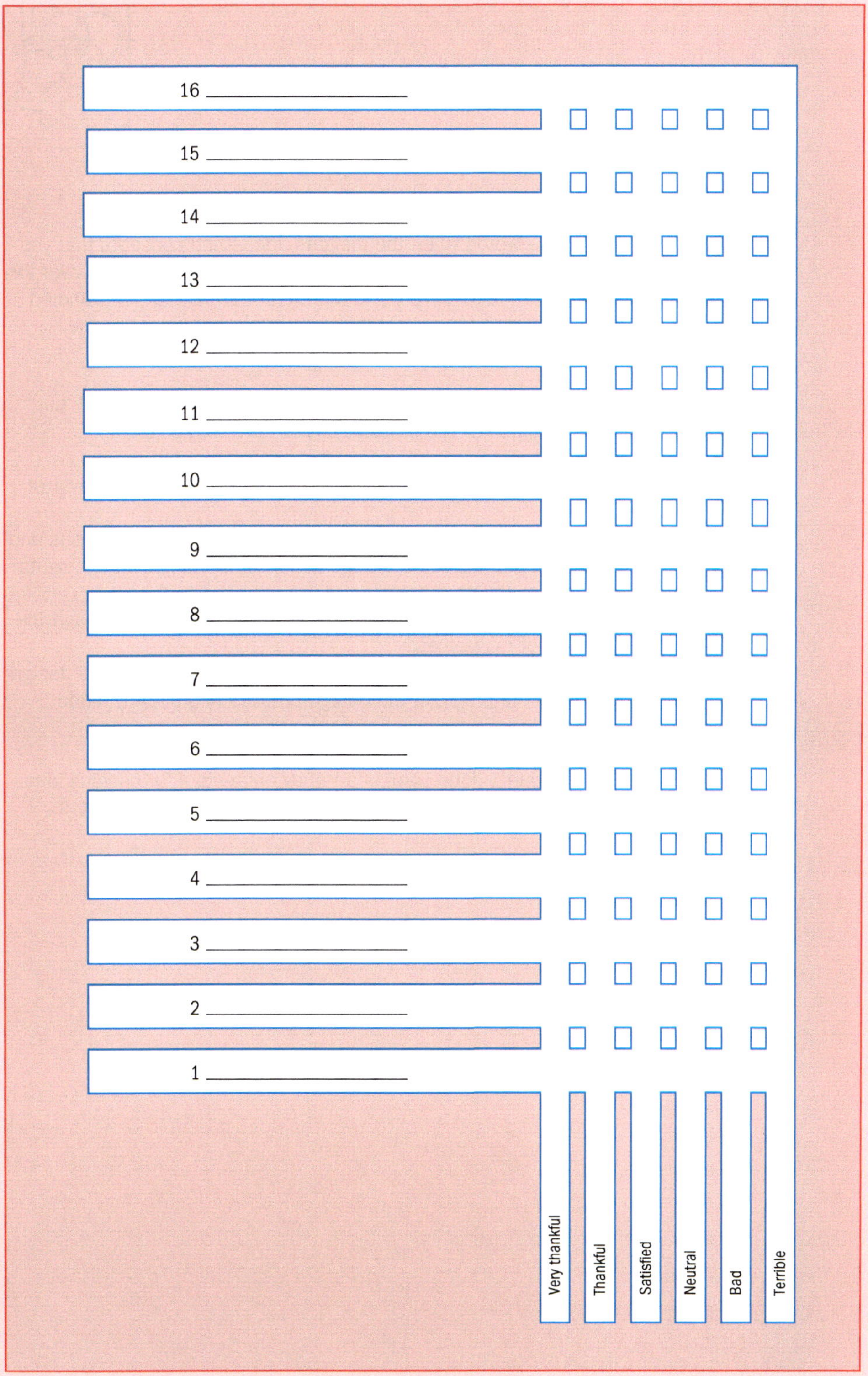

Magazine cover

Fundamental attitude: IMAGINE

What is it?

Why should you use words when the insights, ideas and concepts that you collected can also be shown with images? You can do a magazine cover for clustered insights in the discovery phase, but just as well for the bundled ideas in the definition phase or prototypes in the development phase.

How does it work?

Imagine writing a magazine full of all the insights you have collected, the ideas you have acquired or the prototypes you have worked on:

- Work on the cover of this magazine: what are the most important insights, ideas or prototypes that deserve a place on the front cover of the magazine?
- Choose a title for the magazine, choose photos or other illustrations and come up with titles for articles that contain the most important insights and ideas or describe prototypes.
- Also, provide stopping power: why would a passer-by pick your magazine from a rack full of magazines?
- Make sure you keep the cover handy for constantly reminding the design team how much the progress the design process has already made.

Digital:

- Adobe Spark (spark.adobe.com) is a handy online tool for bringing ideas to life. Adobe Spark also has a feature for making magazine covers.

NOTES

4

Mood board

Fundamental attitude: IMAGINE

What is it?

Whatever stage of the design process you are in, sometimes you want to communicate a certain feeling quickly or you want to look at an idea or a concept from an emotional point of view. Images are more effective at interpreting feelings than words. A mood board is a visualization of a concept, thought or feeling in the form of a collage with different images and texts. The collage makes the goal, the wish, the atmosphere or the concept behind a possible solution clear.

How does it work?

- Physical: collect a pile of magazines and cut out pictures and texts that relate to the feeling that you want to convey.
- Digital: search the internet for pictures and texts and print them out. Use, for example, sources such as:
  - Unsplash (www.unsplash.com)
  - Pinterest (www.pinterest.com)
  - Google Images (images.google.com)
  - Instagram (www.instagram.com)
- Make a mood board by combining images that convey the feeling or get the message across.

NOTES

Observations

Fundamental attitude: EMPATHIZE

What is it?

Within design thinking, research is done through observation to clarify customer demand, see through patterns and to test hypotheses. Observing is a qualitative research method which is about (learning to) look at and record perceptible behavior. It requires you to look carefully and in a structured manner, at how a customer moves around, what he says, what actions he performs, etcetera. Mike Youngblood (2013) describes four techniques that can be used in the design process:

1 Counting occurrences
2 Timing durations
3 Diagramming social interaction
4 Mapping movement

The four techniques are simple, structured and low tech (you only need a pen and paper). This allows you to work quickly and generate data that can be analyzed on the spot. The four techniques are versatile and can be combined with other observation techniques.

How does it work?

1 Counting occurrences

Here, it's all about counting things, people or actions. Youngblood cites an example in which a number of people bring their own bag to the supermarket as opposed to a number of people who buy one in the supermarket. He also explains that there are several variables that can be added to the research. You could also keep track of how many shopping items people are buying in relation to whether or not they are buying a bag.

You proceed as follows:

- Decide what you want to look at and determine the number of variables. The more variables, the more difficult it will be to observe, so limit the number of variables.
- Count the number of times you see the defined variable happen.
- Don't give up: it may be boring, but it yields relevant patterns. Half an hour of counting often already raises questions about the design process. Take the time to formulate these questions. Are there (preliminary) hypotheses for your research? To what extent does that influence further research (for example, in the form of INTERVIEWS)?

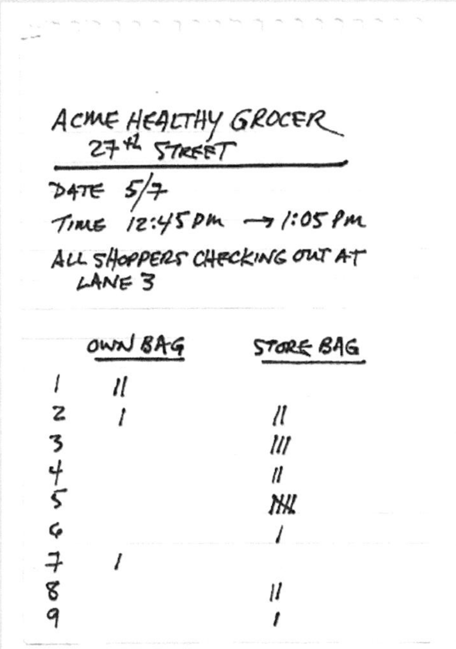

ACME HEALTHY GROCER
27th STREET

DATE 5/7
TIME 12:45 PM → 1:05 PM
ALL SHOPPERS CHECKING OUT AT LANE 3

	OWN BAG	STORE BAG
1	//	
2	/	//
3		///
4		//
5		~~////~~
6		/
7	/	
8		//
9		/

2 Timing durations
Clocking the duration of small events and behavior is used to show how much time a task takes, how much time someone spends on an activity or how long someone exhibits certain behavior. This provides surprisingly different results than you would expect from your observations, predictions and your own estimates.

Youngblood timed durations at a quick-serve restaurant chain. For example, how long does a customer look at a menu before ordering? Is there a difference in time between men and women, elderly and young people or groups and individuals? At a quick-serve restaurant that is reliant on a large volume of customers, it is very interesting to know how to present the menu in such a way that the time between checking out the menu and ordering is minimized.

You proceed as follows:

- Use a watch or stopwatch to keep track of the time.
- Take notes on paper.
- Enter multiple variables for clocking (male/female, young/old, single or with a group).

Use the observations as a solid basis for conducting interviews and for differentiating target groups.

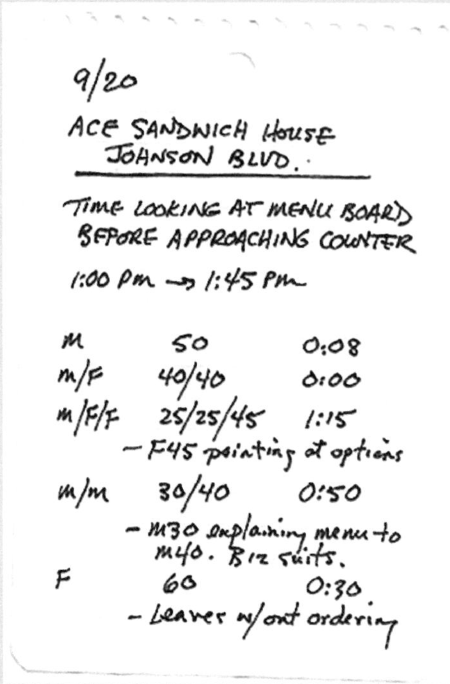
9/20

ACE SANDWICH HOUSE
JOHNSON BLVD.

TIME LOOKING AT MENU BOARD
BEFORE APPROACHING COUNTER

1:00 PM → 1:45 PM

M	50	0:08
M/F	40/40	0:00
M/F/F	25/25/45	1:15
	– F45 pointing at options	
M/M	30/40	0:50
	– M30 explaining menu to M40. Biz suits.	
F	60	0:30
	– Leaves w/out ordering	

3 Diagramming social interaction

Sociograms are used for finding patterns in social interactions or seeing relationships between people. With a sociogram you will use arrows and blocks or circles which show the interactions between people. Youngblood investigated how a seat in a classroom influenced the interaction between students and between students and the teacher.

To investigate this, proceed as follows:

- Draw diagrams with connections between one person and another.
- Keep it simple: only observe a limited number of aspects of one interaction.
- Keep it tidy: the observed do not want to feel being watched or harassed.
- Count how often something occurs:
  - Count how often students in the first, second, third row talk to each other.
  - And when they talk: do they talk about the learning material or do they just ignore the teacher?
- Further research can be done with diagrams:
  - Does the grade a student receive for a course have to do with the student's seat in the classroom?
  - And/or does it have to do with the type of interaction in the classroom?

- You have now turned the diagrams into sociograms. Which conclusions can you draw from the sociograms?
- The results from this example can lead to changes in how the classroom is set up or in limiting (or stimulating) contact between students.

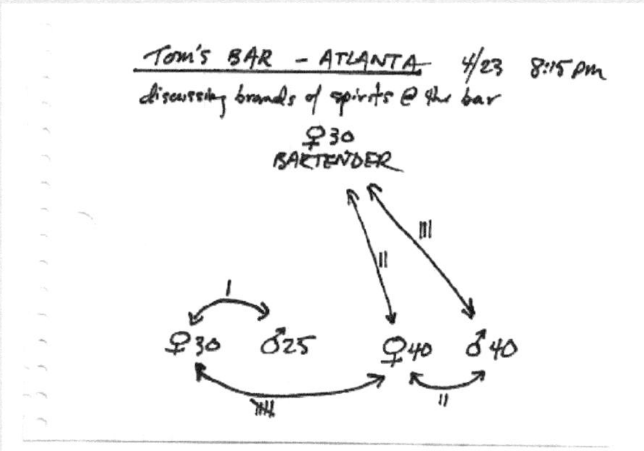

4 Mapping movement

By mapping movements, you can track meaningful motions in physical interactions. It is a technique that you can use for many situations. The only condition is that 'some things are moving, and other things are staying in place'. Youngblood investigated the movement of buyers when viewing a luxury car at a dealership. The results would be used for improving the company's website. What appeared from the observations? Most drivers first sat in the driver's seat but also wanted to know how their co-drivers would like the seats. So they tested several seats. This insight was used to also show features from the perspective of the co-driver on the website.

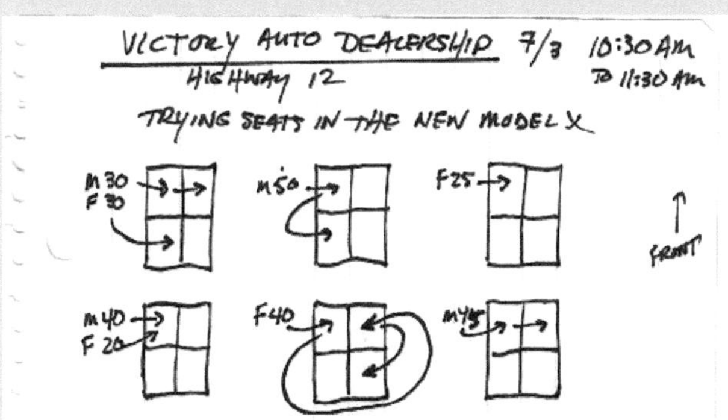

One-hour prototype

Fundamental attitude: EXPERIMENT

What is it?

One-hour prototyping means that the design team can make a representation of the possible solution within an hour. That should be long enough to test the idea or concept and get feedback for a more elaborate prototype. Making a one-hour prototype, also called rapid prototyping, is an effective way to:

- make ideas tangible very quickly;
- learn by doing;
- get feedback from the customer and from stakeholders quickly.

A one-hour prototype can take on various forms: a detailed drawing, a model, a test set-up and a three-dimensional handiwork are just a few examples. A one-hour prototype can also take the form of a STORYBOARD or a MOOD BOARD,

How does it work?

- Determine which form best represents the idea or concept and who the prototype will be shown to.
- Use the materials that are available and only use the available time. It is not about being perfect; just make it good enough to convey the idea.
- Have users and stakeholders test the prototype and note their feedback for the next one-hour prototype.

NOTES

Personas

Fundamental attitude: IMAGINE

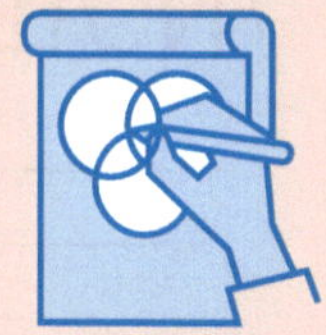

What is it?

Preparing personas helps the design team to identify with future users. Personas are used in design thinking to visualize data from, for example, INTERVIEWS, QUESTIONNAIRES or other sources. Personas are based on factual knowledge and insights into a large range of users who in all likelihood, will behave in the same way. Personas are classified on the basis of character traits, so based on who the users are and not on, for example demographics or purchase volume.

A persona is worked out in detail and is visualized as a specific 'person' with his or her own unique character traits. Based on storytelling, the persona really comes to life. Usually, four to five personas together form a good representation of the total user group. You can also prepare (one or two) non-personas; these are users who are not dealing with the problem or the solution.

When using personas, prevent stereotyping. Stereotyping is based on assumptions, prejudices and generalizations that are not based on collected facts and research. Also keep in mind that personas are meant to be an internal tool for the design team, not as a means to interact and communicate with users.

How does it work?

You can work out the following three steps:

1 Identifying assumptions:
   - Identify assumptions and related stereotypes (in order to be able to break these down later). Identify the typical and traditional user categories for this.
   - Determine what these different users in these traditional categories want to achieve (the user goals). Formulate these goals via 'I want...' or 'I need ...' statements.
   - Write all the 'I want...' and 'I need...' statements, in random order, on separate post-its.
   - Use the post-its to discover patterns in the 'I want...' and 'I need...' statements. The result is a new layout of users sorted in a number of clusters.

2 Form skeletons. Search for 'skeletons' of personas by further detailing the clusters you found:
   - As a design team, discuss the most important needs, objectives and preferences of the different groups that arise. Combine groups that are very similar.

- Make a list of the characteristics per group; these will form the skeletons for the later personas.
- Try to formulate three to five skeletons. Do you have many more? Choose the most important three to five in order to continue.
- Test the skeletons against previously collected insights about the user (for example, from other tools) and adapt them if necessary.

3 Translate skeletons to personas:
- Use the fill in form to make the skeletons come alive in personas.
- Test the personas found with one or more of the EMPATHIZE tools.

You can take further steps in imagining the developed personas:
- Have a life-size persona figure printed with its most important features. Use this when the design team is working on this specific user group.
- Set up persona rooms to bring all the information provided about a persona to life. When you enter this room, it is as if you are inside the world of the persona. Sit in the persona room if you are working for this specific user group.
- Use trained actors (or acting students) to have the design team interact (fictionally) with the personas. You can also use trained actors during the presentation of a prototype. See also: ROLE PLAY.

FILL IN FORM 4.10 Personas

Name: ____________

Age: ____________

Profession: ____________

Status: ____________

Attitude: ____________

Archetype: ____________

[Photo]

How does ____________ relate to ____________

Goals:

What makes ____________ happy?

What makes ____________ unhappy?

Personality
Choose one from the following opposites:

dominant	submissive
passive	dynamic
systematic	chaotic
sensitive	uncertain
idealistic	practical
excitable	cold-blooded
follower	initiator
star	adaptable
seeks crisis	avoids crisis
amiable	formal
aggressive	peaceful

Interests:

Brands:

NOTES

4

Pitching

Fundamental attitude: COOPERATE

What is it?

At different moments in the design process, you would want to convince others of the ideas you have as a design team. This is called pitching. During the discovery phase, these are ideas about the problem situation; during the definition phase, these are ideas about the problem definition; during the development phase, ideas about solutions and during the implementation phase, ideas about the implementing solutions. Use pitching to share your ideas with others.

How does it work?

- Consider which stakeholders you want to convince.
- Do you want to have all stakeholders together or will you organize different meetings for this? Think carefully about the composition of the group before the meeting and approach the stakeholders in accordance with their preference. Some considerations are:
  - Is it essential to get everyone on the same page? Then make sure that everyone involved is present at the pitch.
  - Are people involved who don't get along well? Try to find out why that is and clear the air in advance, or organize different meetings.
  - Is internal coordination not essential? Speak to those involved in the pitch in advance and catch up with them in person.
  - Are there stakeholders who are not interested in your pitch while you consider it important to involve them? Understand the underlying reasons and entice them to come.
- Make sure you inform your guests well in advance. Be clear about the purpose of the pitch:
  - Involve the most important stakeholders when preparing the pitch, for example, by sending them a preview of the presentation and asking them for feedback.
- Attract attention with a good story (for example, tell an anecdote, use humor). Make the pitch visual, but don't overdo it when using slides.
- Be clear about the next steps and what you expect from the stakeholders who were present.

Read more about pitching?

Jeroen van Geel describes a step-by-step plan in his book *Pitching Ideas* (2018) for pitching ideas in different situations. He mentions four pitfalls when pitching an idea and gives tips to avoid them. This concerns the following pitfalls:

1 **Overcompensation mode**. If you do not know what the core of your message is, you run the risk of overcompensating. You will then present an idea down to the smallest detail. Listeners will respond to the details, while they do not know the core message. Tip: present abstract ideas at an abstract level.
2 **Rambo mode**. The idea is not picked up, but you keep repeating the same thing to convince the audience. Tip: stop pitching and pay attention to why your idea doesn't come across.
3 **Best-Idea-Ever mode**. Your belief in the idea is so great that you continue, without investigating the idea any further: in that case, the idea can be shot down but you keep firmly believing in it. Tip: make sure you know everything about your idea, and that you have done enough research, including research on the weak points.
4 **Wrong Style mode**. The focus in your story is wrong because you have mistakenly assumed that others have the exact same motives as you do. Tip: Make sure you always know what your listeners think is important and adapt your story accordingly.

Post-its

Fundamental attitude: THINK FLEXIBLY

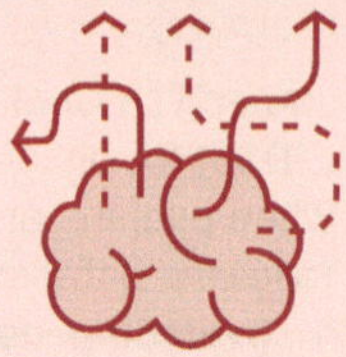

What is it?

At different moments in the design process you may be overwhelmed by a large number of ideas. Then it's time to bring it back to basics. This is possible by writing all ideas on post-its. Working with post-its ensures that you can physically move the ideas around and see links between groups and clusters of ideas a lot faster.

4

How does it work?

- Write or sketch all ideas on separate post-its and stick them on a wall or a table.
- Make one random post-it as your first cluster and go through the other post-its: do they fit in with this cluster or not? This is a matter of intuition, don't think about it too long. If the idea fits in, stick the post-it in the cluster; if it doesn't fit then create a new cluster. Go through all post-its in this way. A maximum of fifteen clusters is manageable, but you will usually end up with fewer clusters. Give the clusters recognizable and stimulating names, for example: 'banana-in-a-bowl, toast, man overboard'.
- Talk to others about the best characteristics of the clusters and combine these with other clusters.
- If necessary, look back at the insights you formulated earlier. Try to combine these with the best elements from the clusters of ideas.
- The resulting new clusters will then form the basis of the next in the design process.
- Determine with your design team, on the basis of, for example, the COCD BOX or the DECISION MATRIX, which clusters will be used to continue the design process.

NOTES

Problem paradox

Fundamental attitude: THINK FLEXIBLY

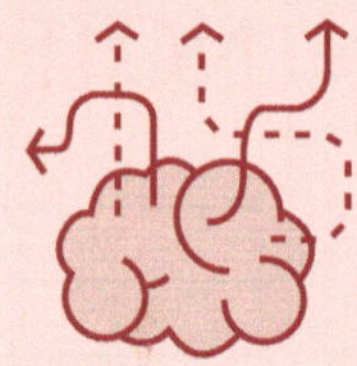

What is it?

A problem paradox is a contradiction within a complex problem, making the initial problem difficult to solve. If you try to resolve a problem paradox, you can get stuck in circular reasoning: every logical remedial action or measure is frustrated by a principle, law or norm.

Kees Dorst (2015) explains a problem paradox based on the 'fitting room problem': to prevent the stealing of clothing (hanging up cameras, monitoring) is hampered by the actual function of the fitting room (the right to privacy).

How does it work?

- Formulating a problem as a paradox helps to gain new insights.
- You can formulate the problem paradox by using the fill in form.
- Complete the circle to expose the paradox of the problem.
- Try to discover what sustains the problem and how the circle can be broken.

Read more about the problem paradox?

Read the book *Frame Innovation: Create New Thinking By Design* (2015) by Kees Dorst.

FILL IN FORM 4.11 Problem paradox

Problem paradox name: ____________________

Because...	...this happens:

Because...	...this happens:

Because...	...this happens:

Because...	...this happens:

Actions to take to investigate the paradox in more detail:

FILL IN FORM 4.12 Full circle

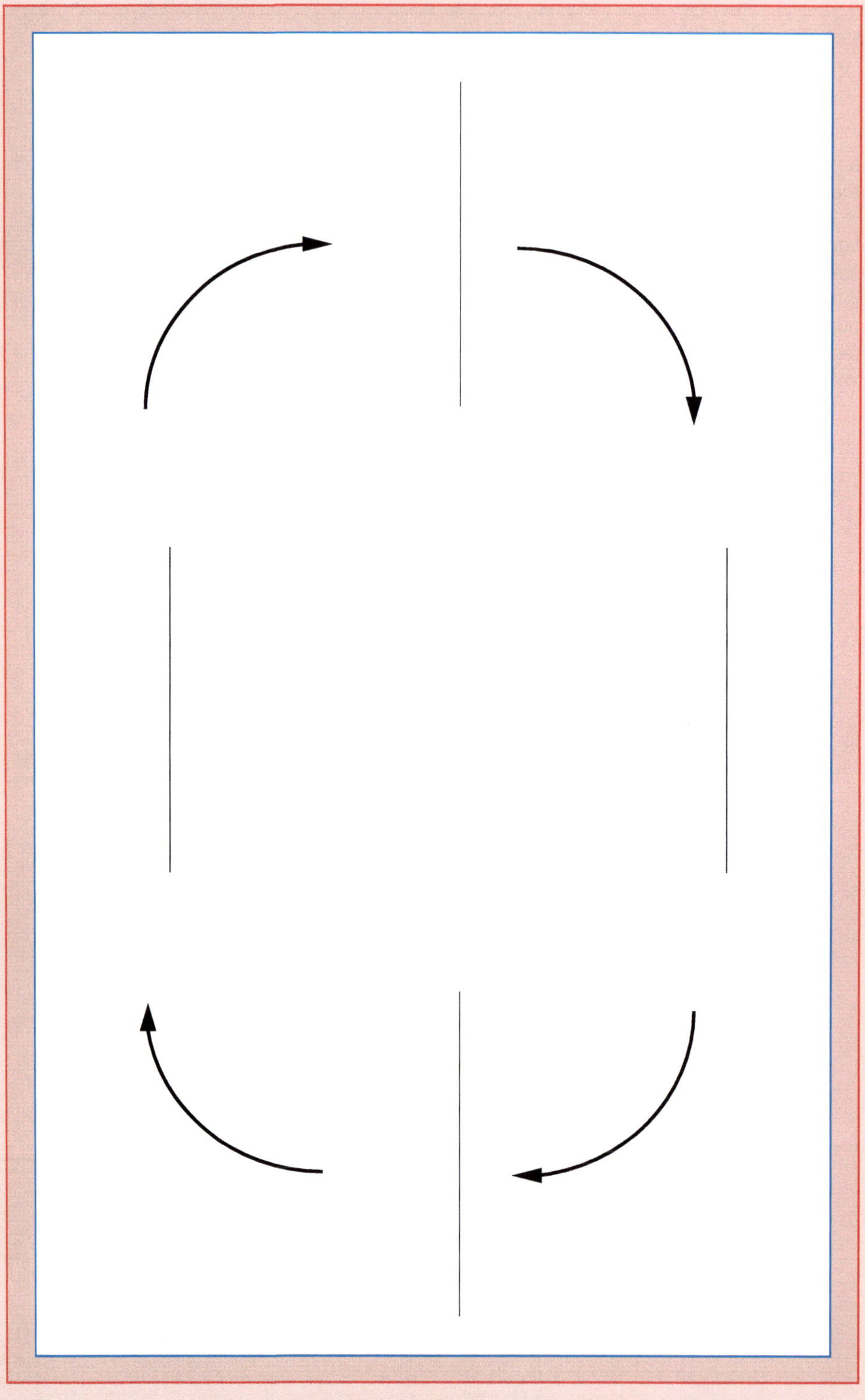

NOTES

4

Questionnaires

Fundamental attitude: EMPATHIZE

4

What is it?

Questionnaires or surveys are an option for gathering a large amount of information. Questionnaires can be filled in via a pollster, individually or digitally. Digital questionnaires do not only save the respondents time compared to physical lists but also save time for the researcher(s). This makes analyzing information faster for use in the design process. To ensure that the correct information is retrieved, the correct questions are essential. Misinterpretation of the questions can lead to incorrect answers. Whereas INTERVIEWS during a design process are about the unique conversation and the unique insights of the interviewee, with questionnaires you strive for a certain number of responses so that the information is representative of an entire population. Working with a survey or questionnaire requires thorough preparation.

How does it work?

To prepare:

- Consider the purpose of the questionnaire:
  - What do you want to achieve?
  - Which questions do you want to have answered or which assumptions are to be tested?
- Consider who you want to question.
- Do you want to have the questionnaire completed digitally or physically? This depends on the users you have defined.
  - For physical collection: with a pollster or individually?
- Make sure it takes no more than five minutes to complete the questionnaire; this increases the chance of response. Does it take (much) longer to complete because you need more information? Then make two parts and give each user just one of the two parts to complete.
- Provide a clear introduction that describes:
  - the purpose of the questionnaire;
  - approximately how long it will take to complete;
  - how the collected data is handled and how confidentiality is guaranteed.
- Do not send the questionnaire to all users immediately. Start with ten people and add the question: ‘Do you have any comments about this questionnaire?’ With the feedback you get, you can improve the questionnaire before sending it to other users.

Questions:
- Place the questions in a logical order, viewed from the respondent's perspective. Ask the most interesting questions for the (future) user first, to stimulate interest.
- Be specific and prevent the use of professional terminology, jargon or catch-alls:
  - not: 'What do you think is innovative about our product?'
  - but: 'Which aspects of our product have you not seen anywhere else?'
- Avoid double negatives:
  - not: 'At what point in our service were you not dissatisfied?'
  - but: 'At what point in our service were you satisfied?'
- Ask one thing per question:
  - not: 'When was the last time you were in a hospital and how did you feel then? '
  - but: 'When was the last time you were in a hospital?' And 'How did you feel the last time you were in a hospital? '
- Do not try to steer the questions. The user will then provide information you want:
  - not: 'Did you find it uncomfortable to talk to the dean about your problems?'
  - but: 'How did you feel when you talked to the dean about your problems?'
- Don't jump right in but introduce questions by giving the context:
  - not: 'What did you think of our welcome email?'
  - but: 'New customers always receive a welcome email from us after their first order. What did you think of this email?'

To increase response:
- Give a reward or prize to (future) users when the questionnaire is fully completed.
- Stay close to the respondent (if possible). It is okay to be a little compelling: 'Will you fill in the questionnaire immediately?' or: 'I will come to retrieve the questionnaire in five minutes.'
- Is response lacking? Send a reminder after three to four days.

If you are looking online for ways to send a questionnaire, look at these:
- Google Forms (docs.google.com/forms)
- SurveyMonkey (en.surveymonkey.com)
- Typeform (www.typeform.com)

4

Quotation process

Fundamental attitude: COOPERATE

4

What is it?

Many organizations do not employ design thinkers, but get their in-house design expertise by hiring an external design thinker. Which means that there is a quotation process prior to the start of a design project. The potential external design thinker will invest time in preparing and creating a DESIGN PROJECT ROADMAP.

How does it work?

- The quotation process consists of a debrief of the quotation meeting. The client would want to see his input in this report:
  - the question asked by the client;
  - the client's expectation about the results;
  - what has been done about the design problem before;
  - what background information is available;
  - which lead time the client has in mind;
  - what budget the client has in mind for the project or what budget is available.
- When preparing the quotation, questions may arise that have not been discussed or that are not in the scope of the assignment. Ask if the client can answer these questions immediately. In doing this, the client can fine-tune the assignment and the quotation becomes a joint effort, which can increase commitment to the project.
- Quotations that are sent by post or email, can start to lead a life of their own. Always discuss the proposal in person, so that misinterpretations can be straightened out immediately. It is also possible to (re)write the proposal together with the client. Here too, co-creation increases the chance of success.
- Without support, a project is doomed to fail. The client is not the only one whom the design team depends on; there are all sorts of other people involved who must be convinced of the importance of the project. Determine together with the client, who the most important stakeholders are (in this preliminary phase of the design process) and which arguments would create support for the project.
- Clarify the expectations and the role of the client and set explicit conditions.

NOTES

Removal box

Fundamental attitude: WORK INTEGRALLY

What is it?

If already you know about the history of a project in the discovery phase of the design process, (which information is already available and the earlier attempts to solve the problem), then you do not start the process somewhere randomly. Make the history of a project tangible by collecting information in a removal box (or shoebox) and let the collected information inspire you during the discovery phase. Nowadays, a lot of information is digital and you would want a digital removal box. Remember that the power of imagination is what makes information tangible and ensures that it retains a lot better than a folder in one shared Dropbox or on Google Drive!

How does it work?

- Take the removal box during the step 'take a dive in the past' everywhere you go. Collect all (tangible) information in the box about the past of the project or problem. Think of:
  - research reports
  - evaluations
  - tangible results (also think of visualizations made)
  - memos and interview reports
- Try to find out what inspired the previous project team. For example, which books and literature did they read?
- Use the removal box for important milestones and ask yourself: Have we forgotten important lessons from the past because of our enthusiasm for our own ideas? Do we have the tendency to making the same mistakes?

NOTES

Roadshow

Fundamental attitude: COOPERATE

4

What is it?

It is important to involve others during the design process. This can be done via a roadshow in which you actively seek out those involved to catch up, based on visualizations.

How does it work?

- Provide visualizations of the steps that have been followed so far in the design process.
- Invite those involved personally. Ask them to register in advance, that is how people will commit to coming.
- Set up a room with the visualizations and communicate when and at which times you will be present to explain the design process.
- Talk to stakeholders about the intended direction of the design process.

Alternative: eat a croissant together

- Plan a number of mornings on which you schedule half an hour for tea or coffee with a croissant (or muffin, banana, sandwich, etcetera) to inform stakeholders about the design process.

NOTES

Role fulfillment

Fundamental attitude: COOPERATE

What is it?

Make the team members jointly responsible for the design process by assigning roles at the start of the design project. Think about the following four roles:

- The **team captain** is the cooperative foreman in the process who monitors and ensures that the team is making progress. In discussions, the team captain monitors the time schedule and the results that are to be achieved.
- The **ambassador of the customer** constantly encourages the team to look at the process through the customer's eyes. By asking the right questions, the ambassador of the customer ensures that the end-user plays a central role in the whole design process.
- The **challenger** ensures that the team stays focused on the market opportunities and technical feasibility. With the proper questions, the challenger makes sure that the project team is really innovative in its solutions, without losing sight of reality.
- The **connector** ensures the correct connections are made between the team, internal organization and external stakeholders. The connector also searches constantly for the proper links with other project teams or other projects that are happening within the organization.

How does it work?

The question cards offer an aid for filling in the various roles during the entire design project. Use the question cards and complete the questions that are important for your specific role.

- Have you determined other team roles? Compose question cards for these specific roles with the design team.
- Print the cards and take them to team meetings. That way, yourself and others stay keen and focussed on the role that everyone plays within the design team.

FILL IN FORM 4.15 Role fulfillment

Team captain:

Is the time path/planning still correct?
Do we have everything we need to attain the next milestone?
What will we agree on for the next meeting?
Can we make concrete appointments for the next meeting?

___
___
___
___

Ambassador of the customer:

What do the different personas think - based on their personality traits?
Will the customer really be helped with this solution?
Is there (customer-related) research which supports this opinion?
What does the customer really need?

___
___
___

Challenger:

How do others do it?
Are there already other examples or alternatives?
Is it feasible in terms of cost?
Is it technically feasible?

___
___
___
___

Connector:

Who do we need for this?
Who must we inform about this?
Does person 'x' know about this?
Which examples of projects and teams can inspire and motivate us?

___
___
___

4

Role play

Fundamental attitude: EXPERIMENT

What is it?

A role play is a simple variant of the DESKTOP WALKTHROUGH. It is a way to experience a product or a service from the perception of the user. A physical experience is played out in real life during a role play to bring out the weak spots of a product or service. See also CUSTOMER JOURNEY MAP. Improvements can be made to the product or service on the basis of a role play.

How does it work?

1 Define and replay:
- Determine which script fits the problem you want to solve or the solution you want to test.
- Write your script. Use, for example, a JOURNEY OF EMOTIONS, a STORYBOARD or CUSTOMER JOURNEY MAP for this.
- Give different people a role and write the roles on a card. Give them some time to study their role, for example, with 10 MINUTE GOOGLE, and answer any questions they may have.
- If necessary, set up a room where the role play takes place.
- Replay the devised script and allow people to think out loud.

2 Draw conclusions. Draw conclusions after replaying the experience:
- What can you learn from this?
- What are the 'mistakes' in the experience?
- What new insights have you come to?
- What new ideas did you get?

NOTES

4

Rules of thumb

Fundamental attitude: EXPERIMENT

What is it?

In design thinking, solutions or prototypes are ideally tested by users. If this is not possible due to lack of time or resources or, if you want to pre-test, then solutions and prototypes can also be tested based on a number of rules of thumb. By rules of thumb, we mean insights from previous scientific or practical research. In other words, rules that have already been discovered.

How does it work?

- Determine which solution or prototypes you cannot test with users.
- Search for insights from previous research that are applicable.
- Search the online database for research at your school or use websites like scholar.google.com or academia.edu.
- As a design team, determine the rules of thumb that you want to use for the solution or for testing the prototype.
- Use the insights to test and improve the prototype.

NOTES

Scenarios

Fundamental attitude: WORK INTEGRALLY

What is it?

With the help of scenarios, you can investigate how the solutions that have been found for a problem can be improved - by reasoning how the solutions would function in the future.

How does it work?

- Per user-group (expressed in PERSONAS or CHARACTER PROFILES), work out one of the possible solutions in two scenarios, as follows:
  1 a scenario in which the user group is very positive about the solution.
  2 a scenario in which the user group is very negative about the solution.
- Describe - step by step - what happens per user-group for each scenario while experiencing the possible solution. If possible, use the fill in form for the CUSTOMER JOURNEY MAP or JOURNEY OF EMOTIONS.
- Use the new insights from the scenarios to formulate the problem or an idea for a solution.
- Work out the positive and negative scenarios based on the improved solution.
- Repeat this process until the solution is optimized.

NOTES

Shadowing

Fundamental attitude: EMPATHIZE

What is it?
Shadowing is a variant of OBSERVATION. While observation is all about observing a group of users, with shadowing you focus on just one (future) user. The purpose of shadowing is to gain insight into someone's behavior. You keep your eye on the target while noting down all your observations.

How does it work?
- Determine in advance which behavior you will explicitly pay attention to:
  - What body language does the user have?
  - How does the user react to other users?
  - How does the user interact with others?
  - Which routines are used more often in the user's behavior?
- If you have already prepared a CUSTOMER JOURNEY MAP for the (future) user, then use it as a guide for shadowing. Evaluate whether your observations match the steps in the customer journey.

NOTES

4

Spiderweb

Fundamental attitude: WORK INTEGRALLY

4

What is it?
Stakeholders are people or organizations that are interested in, or influence the design process. Think of future users, involved interest groups and managers who use the project to put themselves on the agenda, and so on. A stakeholder can be a positive or negative influence on the design process or they can be positively or negatively influenced by it themselves.

Using a spiderweb you can identify the most important people involved in a problem or subject quickly and show the interconnections.

How does it work?
- Note down the problem or subject in the middle of the form.
- Around this, write down the five most important stakeholders (first layer).
- Write down five more stakeholders per stakeholder that have influence in one way or another (second layer).
- Try to discover connections: who knows whom and how many more people are involved that can influence your design process?
- Consider which stakeholders you should involve (more) in the design process and how you want to do that.

FILL IN FORM 4.16 Spiderweb

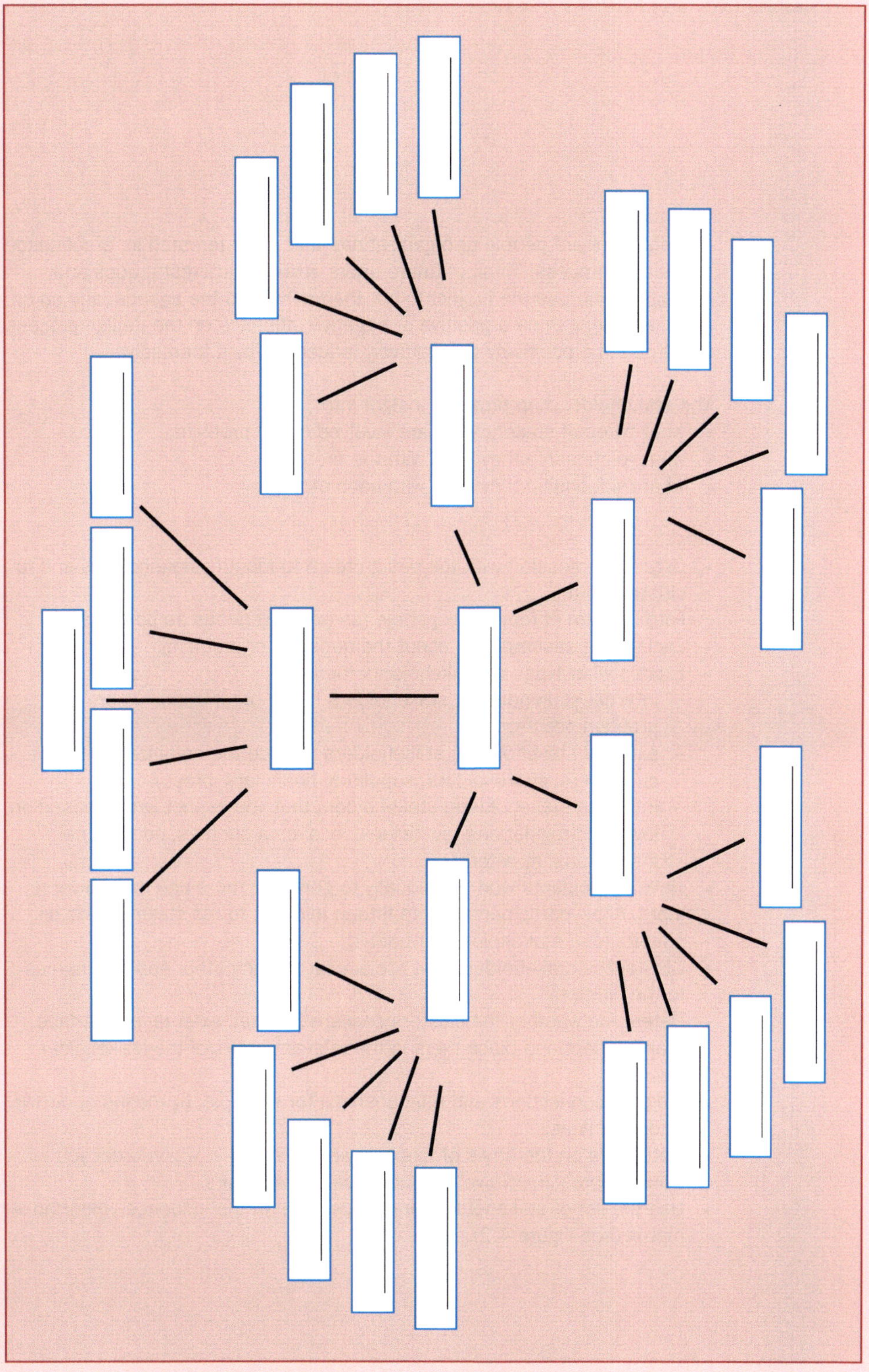

Stakeholder map

Fundamental attitude: WORK INTEGRALLY

What is it?

Stakeholders are people or organizations that are interested in, or influence the design process. Think of future users, involved interest groups and managers who use the project to put themselves on the agenda, and so on. A stakeholder can be a positive or negative influence on the design process or they can be positively or negatively influenced by it themselves.

The stakeholder map provides insight into:

- what types of stakeholders are involved in the problem;
- what relationship they have with the problem;
- what relationship they have with each other.

How does it work?

- Organize a session with the design team to identify stakeholders and to cluster them.
  Form a team of four to six people (as heterogeneous as possible).
- Explain the assumptions about the question or problem.
- Explain what types of stakeholders there are:
  - internal stakeholders: stakeholders in the organization where the problem occurs;
  - external stakeholders: stakeholders outside the organization (users, customers, shareholders, suppliers, financiers, press):
  - interface stakeholders: stakeholders that you cannot ignore based on laws and regulations (politicians, local communities, educational institutions, et cetera).
- Have participants work individually to generate more and more diverse input. Allow participants ten to fifteen minutes to list stakeholders on post-its (one stakeholder per post-it).
- Cluster the stakeholders that are similar to each other and give names to the clusters.
- Determine whether the clusters concern internal, external or interface stakeholders and place them in the relevant areas of the stakeholder map.
- Indicate connections and relationships, for example, by means of arrows or dotted lines.
- Determine on the basis of this overview how and in which order you would approach, involve and influence stakeholders.
- Use the list of stakeholders and place them in the influence/importance matrix (see Figure 4.2).

FIGURE 4.2 Influence and importance matrix

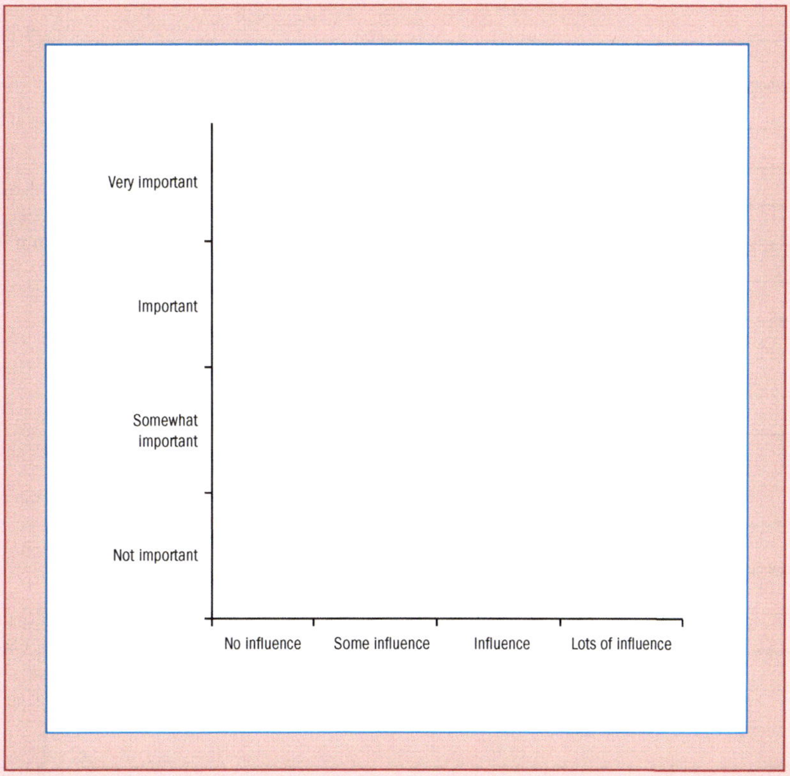

The next step consists of involving some of the most important stakeholders. Organize a session with them to complete the current stakeholder map by asking them:

- What relationship do you have with the problem? How does this influence you (in a positive or negative sense)?
- What challenges do you think there are?
- Where would you position yourself? Use an empty stakeholder map for this.
- Which stakeholders are more involved according to you (and what is your relationship with them)?

FILL IN FORM 4.17 Stakeholder map

Internal stakeholders:

External stakeholders:

Interface stakeholders:

NOTES

4

Storyboard

Fundamental attitude: IMAGINE

What is it?

A storyboard originally comes from the motion picture world. It is a series of drawn shots from a movie script. By preparing a storyboard you visualize the experience of a user based on different shots.

How does it work?

- Use the developed PERSONAS or CHARACTER PROFILES as a starting point.
- Think of different shots. Tomitsch et al. (2018) use the following types of shots:
  - *wide shot*: shows the environment or context;
  - *long shot*: shows a person or thing in the environment or context;
  - *medium shot*: shows a person's head and shoulders;
  - *over-the-shoulder shot*: shows a thing or person 'over the shoulder' of another person;
  - *point of view shot*: show things through the eyes of the person concerned
  - *close-up shot*: gives a detailed picture of something or someone.
- Also consider the use of, for example, speech bubbles, various colors and arrows to represent the user's experience.
- Draw the different shots on the form.

FILL IN FORM 4.18 Storyboard

1	2	3	4
5	6	7	8
9	10	11	12
13	14	15	16

Team meeting

Fundamental attitude: COOPERATE

4

What is it?

Team meetings are arranged regularly during the design thinking process and are an important point of contact between design team members. Prepare these meetings well and arrange for a team captain to supervise the meeting. In this way, the meetings will really help to inspire and move the design process forward.

How does it work?

- Use a fixed day/time in the week, so that team members will be able to plan their own work around it.
- Planning fixed appointments with team members with a busy schedule is sometimes difficult. Plan the meetings around lunchtime (11:30 am to 1:00 pm) or at the end of the day (5:00 to 6:30 pm). Provide a simple meal and make your team members happy.
- Keep meetings short, one hour at most. With a time limit, people are more focused and meetings run more efficiently.
- Stick to the rule: whoever is there is there. Don't reschedule the meeting because not everyone can make it.
- Start each meeting with a stand-up: each team member stands up to give a one-minute (not longer) progress report.
- Conclude every meeting with a quick check if every team member can move forward and every team member is satisfied and positive.
- Does a scheduled meeting not seem to be of use? In that case, skip it.

NOTES

4

User diaries

Fundamental attitude: EMPATHIZE

What is it?
User diaries provide insight into patterns in the behavior of users. Have (future) users of a product or service make photos, video's or sound recordings, in the form of a diary, to gain insight into their lives in general.

How does it work?

- Give (future) users a diary and ask them to write down what they do on a daily basis. Ask (future) users to take photos, videos or sound recordings with their phones of important events. Usually, a week of this will give enough information.
- Allow future users to describe concrete circumstances and activities that are related to important aspects of their lives.
- Ask no or few questions, so as not to influence the (future) user or to make them aware of what they do in their day-to-day activities.
- If you nevertheless ask questions, for example by writing them down at particular dates in the provided diary, ask open questions in simple language.
- Analyze the diaries for patterns and insights.
- If necessary, organize a concluding INTERVIEW to check insights or by asking about ambiguities.

NOTES

Watchtower

Fundamental attitude: WORK INTEGRALLY

4

What is it?

During the discovery phase, the watchtower is used to determine what has already been done outside the organization and what is currently happening in area of the possible design problem.

How does it work?

Focus your attention on the outside world by answering questions. Use the form to record your most important findings. Do you cooperate with others? Then use BRAINSTORMING to complete the form.

FILL IN FORM 4.19 Watchtower

Subject: ____________________

Which comparable projects are there in the world?

Which inspiring products or services relate to the problem?

Is there ongoing debate about the problem? If so, what are the pro's and con's?

Who are the inspiring persons or companies relating to the problem?

Who are the competitors according to the client?

Are there any interesting start-ups tackling a different problem or challenge?

Who are the competitors from the users' perspective?

What would be game changing in the business/branch where the problem prevails?

Source: leanservicecreation.com

Workshop

Fundamental attitude: COOPERATE

What is it?

A workshop is a session in which team members, users and stakeholders get involved in the design process. You can organize various workshops during the entire design process.

How does it work?

- When organizing a workshop, consider the objectives:

As the organizer of the workshop, what information do you want to see collected at the end? What results do you want to have achieved?

- Consider who should be invited to reach your goals. Think in terms of number of people, personal characteristics, mutual relationships and attitude towards the design team or towards the subject of the workshop.
- Determine which atmosphere suits the target group and the goals you are setting.
- Determine the theme of the workshop and possibly give it a promotional title.
- Determine the global structure. Think about the introduction, the storyline, tools, intermezzo's, surprises and a good conclusion. Also, do not forget the follow-up to any results.
- Think about what participants can do before the workshop and choose the right communication channel to report this to them (physically, orally, by telephone, via email, by letter).
- Plan the activities that you want to perform.
- Consider what needs to be arranged in practical terms: a good space, materials, a snack and drink.

Need inspiration for different forms you can use in workshops?

Read *The Large Working Book 2* (in Dutch: *Het Groot Werkvormenboek 2*) 2 (2017) by Sasja Dirkse-Hulscher and Angela Talen.

NOTES

Literature references

Andel, P. & Brands, W. (2014). *Serendipiteit, De Ongezochte Vondst*. Amsterdam: Nieuw Amsterdam Uitgevers.

Binkhorst, E., Dekker, T. den & Melkert, M. (2010). Blurring boundaries is cultural tourism research. In: G. Richards & W. Munsters (Eds.), *Cultural Tourism Research Methods* (p. 41–51). Wallingford: CAB International.

Blomkvist, J., Fjuk, A. & Sayapina, V. (2016). *Low Threshold Service Design: Desktop Walkthrough*. Paper presented at: ServDes 2016, Kopenhagen.

Brenner, W., Uebernickel, F. & Abrell, T. (2016). Design Thinking as Mindset, Process, and methods box. In: W. Brenner & F. Uebernickel (Eds.), *Design Thinking for Innovation*, Research and Practice (p. 3–21). Cham: Springer International Publishing Switzerland.

Brown, T. (2008). *Design Thinking*. Harvard Business Review, 86(6), 85–92.

Brown, T. (2009). *Change by Design*. New York (NY): HarperCollins Publishers.

Dekker, T. den (2019). Designing Cultural Tourism Experience. In: D.A. Jelinčić en Y. Mansfeld (Eds), *Creating and Managing Experiences in Cultural Tourism*. Singapore: World Scientific.

Design Council (2005). *The Design Process*. Accessed: https://www.designcouncil.org.uk/news-opinion/design-process-what-double-diamond.

Dobelli, R. (2015). *De kunst van het helder denken*: 52 denkfouten die je beter aan anderen kunt overlaten. Amsterdam: De Bezige Bij.

Doorley, S., Holcomb, S., Klebahn, P., Segovia, K. & Utley, J. (2018). *Design Thinking Bootleg*. Stanford (CA): Stanford University.

Dorst, K. (2015). *Frame Innovation: Create New Thinking By Design*. Cambridge: The MIT Press.

Futurice (2016). *Lean Service Creation*. Helsinki: Futurice.

Geel, J. van (2018). *Pitching Ideas*. Amsterdam: Bis Publishers.

Gloppen, J. (2009). Perspectives on Design Leadership and Design Thinking and How They Relate to European Service Industries. *Design Management Journal*, 4(1), 33–47.

Gorodsky, J. & Rubin, P. (2014). In: Kelley, T. & Kelley, D. (2014), *Creative Confidence* (p. 190–191). London: William Collins.

Herfurth, L. (2016). *Organisations as Artefacts*. An Inquiry into Hidden Design Activities Within Situated Organisational Contexts (Dissertatie). Lancaster: Lancaster Institute for Contemporary Arts.

Highmore Sims, N. (2007). *Workshops*. Amsterdam: Pearson Education Benelux.

IDEO (n.d.). *Project mood sheet*. Accessed: http://hackthesystem.com/blog/habits-failure-and-the-creative-process-how-ideo-the-worlds-premier-design-firmsucceeds-by-expecting-failure/

Kelley, T. & Kelley, D. (2014). *Creative Confidence*. London: William Collins.

Khurana, A. & Rosenthal, S.R. (1998). Towards Holistic 'Front Ends' In New Product Development. *Journal of Product Innovation Management*, 15, 57–74.

Kimbell, L. (2014). *The Service Innovation Handbook*. Amsterdam: BIS Publishers.

Kolko, J. (2010). *Abductive Thinking and Sensemaking*: The Drivers of Design Synthesis. Massachusetts Institute of Technology Design Issues, 26(1), 15–28.

Kolko, J. (2015). Design Thinking Comes of Age. *Harvard Business Review*, 93(9), 66–71.

Kosara, R. (2007). *Visualization Criticism – The Missing Link Between Information Visualization and Art*. Artikel gepresenteerd op: International Conference Information Visualization 2007, Zürich.
Kouprie, M. & Sleeswijk Visser, F. (2009). A framework for empathy in design: stepping into and out of the user's life. *Journal of Engineering Design*, 20(5), 437–448.
Martins, R. (2016). Increasing the Success of Service Design Implementation. *Touchpoint Magazine*, 8(2), 12–14.
Menold. J., Jablokow, K. & Simpson, T. (2017). Prototype for X (PFX): A holistic framework for structuring prototyping methods to support engineering design. *Design Studios*, 50, 70–112.
Newman, D. (n.d). *The design squiggle*. Accessed: https://cargocollective.com/central/The-Design-Squiggle
Nijs, D. & Peters, F. (2002). Imagineering. Amsterdam: Boom uitgevers.
Norman, D. (2013). *The Design of Everyday Things*. New York (NY): Basic Books.
Ostenwalder, A. & Pigneur, Y. (2009). *Business Model Generation*. (s.n.): Alexander Ostenwalder & Yves Pigneur.
Osterwalder, A., Pigneur, Y., Bernarda, G. & Smith, A. (2014). *Value Proposition Design*. Hoboken (NJ): John Wiley & Sons.
Prabir, S. & Chakrabarti, A. (2017). A Model for the Process of Idea Generation. *The Design Journal,* 20(2), 239–257.
Rittel, H.W.J. & Webber, M.M. (1973). Dilemmas in a General Theory of Planning. *Policy Sciences*, 4(2), 155–169.
Simonton, D.K. (1999). *Origins of Genius: Darwinian Perspectives on Creativity*. New York: Oxford University Press Inc.
Stickdorn, M., Hormess, M., Lawrence, A. & Schneider, J. (2018). *This is service design doing*. Sebastopol (CA): O'Reilly Media, Inc.
Stompff, G. (2018). *Design Thinking: Radicaal veranderen in kleine stappen*. Amsterdam: Boom uitgevers.
So, C. & Joo, J. (2017). Does a Persona Improve Creativity? *The Design Journal*, 20(4), 459–475.
Tomitsch, M., Wrigley, C., Borthwick, M., Ahmadpour, N., Frawley, J., Kocaballi, A.B., Núnez-Pacheco, C., Straker, K., Loke, L. (2018) *Design. Think. Make. Break. Repeat*. Amsterdam: BIS Publishers.
Travers, A. (2013). *A Pocket Guide to Interviewing for Research*. Penarth: Five Simple Steps.
Yoo, Y. & Kim, K. (2015). How Samsung Became a Design Powerhouse. *Harvard Business Review*, 93(9), 73–78.
Youngblood, M. (2013). Four Bedrock techniques for observational research. *QRCA Views*, 12(2), 28–35.

Notes per chapter

In addition to the books and magazines from the reference list, there are more sources that were consulted. An overview is included here for each chapter.

Design thinking is ...

Harvard Business Review: The edition referred to here is the USA September 2015 edition of Harvard Business Review, volume 93, number 3, the article by Jon Kolko is called *Design Thinking Comes of Age*.

Quote Geoff Cubitt: Geoff Cubitt is CEO of digital agency Isobar U.S; the quote comes from https://www.garicruze.com/the-blog/?offset=1397402350079

Current complex problems or issues are connected to other problems (interdependent, cross-organizational and dynamic): Kees Dorst writes about the characteristics of complex problems in his book *Frame Innovation*.

'**Loose wild game' text box**: This text is based on H.W.J. Rittel and M.M. Webber (1973). Dilemmas in a General Theory of Planning. *Policy Sciences*, 4 (2) and on a blog about the book '*Wicked Problems*: Problems Worth Solving' by Jon Kolko, https://ssir.org/articles/entry/wicked_problems_problems_worth_solving.

Minors in the field of design thinking: This is a selection offered by Universities of Applied Sciences, found at https://www.kiesopmaat.com

Design based education: Named by one of the co-readers, teacher Froukje van Houten, more information about this can be found at: https://www.stendenblogs.com/klaas-wybo-vander-hoek/blogs/design-based-education-dbe-het-onderwijs-concept-van-nhl-stenden

Case 'What Design Can Do': The idea for this case came from reading the article 'Refugee Challenge is Trumpian overestimation' by Jeroen Junte from the newspaper *de Volkskrant* of 28 June 2016, https://www.volkskrant.nl/columns-opinie/refugee-challenge-is-trumpiaanse-overschatting~bb5806e4/; additional information is found on the website What Design Can Do, https://www.whatdesigncando.com/refugeechallenge/.

Tim Brown's statement comes from the TEDtalk 'Tim Brown urges designers to think big', available at https://www.ted.com/talks/tim_brown_urges_designers_to_think_big.

1 Design thinking is a way of thinking

Opening case 'Vera Winthagen brings happiness to the neighborhood': This article was written by Ilse Zeemeijer and published in *Het Financieele Dagblad* on October 19, 2017, https://fd.nl/morgen/1222771/vera-winthagen-brengt-geluk-in-de-wijk.

Symbols for the six fundamental attitudes: Edited symbols from the Noun Project, thenounproject.com

Think flexibly: This is partly based on the insights Tim Brown provides in his book *Change by Design*, specifically on pages 66 to 68.

Balancing between zooming in and zooming out: The reference to Rosabeth Moss Kanter comes from a MIT / Sloan Management Review blog, Leslie Brakow of March 9, 2011, https://sloanreview.mit.edu/article/rosabeth-moss-edgers-zoom-in-zoom-out-metaphor.

Assignment 'What stands out?': The idea for this exercise is inspired by the photo analyzes that Hans Aarsman previously did every week in Dutch newspaper *de Volkskrant*.

Quote Ad van Berlo: This is from an interview by Ilse Zeemeijer, published in *Het Fnancieele Dagblad* on March 19, 2016, https://fd.nl/morgen/1143633/design-thinking-helpt-om-in-de-toekomst-te-kijken.

Assignment 'Yes, but...': The exercise was conducted with the audience of the Service Design Days 2016 in Barcelona, October 6 and 7, 2016.

T-shaped persons: These are named by Tim Brown in his book *Change by Design* on page 27 and Andy Boyton published a blog on October 18, 2011 about this on the Forbes Magazine website, https:/www.forbes.com/sites/andyboynton/2011/10/18/are-you-an-i-or-at/#295954d16e88.

Innovation sweet spot: feasibility, viability and desirability: These concepts are named by Tim Brown in his article for *Harvard Business Review* from 2008: 'Design Thinking'; and elaborated in a blog on Medium.com by Kristann Orton, published on March 29, 2017, https://medium.com/innovation-sweet-spot/desirability-feasibility-viability-the-sweet-spot-for-innovation-d7946de2183c. The image is an adaptation of IDEO U.

Four phases in developing empathic capacity: The image is an edit of Kouprie et al. (2009).

Newspaper clipping 'Student in the world of the patient': This article appeared on October 18, 2018 in *Dagblad de Limburger* and was written by Hennie Jeuken, https://www.limburger.nl/cnt/dmf20181018_00077842/student-in-wereld-van-patient.

Henry Ford Quote (probably): Although widely accepted, Patrick Vlaskovits describes in a blog for *Harvard Business Review* that there are no 'hard facts' to prove that Henry Ford actually made this statement, https://hbr.org/2011/08/henry-ford-never-saidthe-fast.

Quote Steve Jobs: The *Wall Street Journal* provides an overview of quotes by Steve Jobs at https://blogs.wsj.com/digits/2011/08/24/steve-jobssbest-quotes and indicates that this quote comes from an interview that BusinessWeek did with Steve Jobs, published on May 25, 1998.

'**You will never understand' text box**: This text is inspired by an article on the website *Woman in the World*, in which the commercial is discussed, https://womenintheworld.com/2016/10/22/malepoliticians-in-japan-get-pregnant-in-order-to develop-empathy-with-overworked-mothers/.

Google's investigation into successful teams: described by Richard Feloni on November 18, 2016 in an article for *Business Insider UK*, https://www.businessinsider.com/google-explains-top-traits-of-its-best-teams-2015-11?international=true&r=US&IR=T; the entire investigation can be found at https://rework.withgoogle.com/guides/understanding-team-effectiveness/steps/introduction/.

Quote Inge Rijnders: Inge Rijnders, who works at Zuyd Hogeschool, was proofreader of this book; the quote is from one of her comments on the section about the fundamental attitude COOPERATE.

Interdisciplinary versus multidisciplinary: This text is partly based on a blog post at https://www.startpagina.nl/v/wetenschap/vraag/397316/verschil-tussen-multidisciplinair-interdisciplinair-transdisciplinair/.

Learning to cooperate in a team: Gorodsky and Rubin are mentioned on pages 190 and 191 in the book *Creative Confidence* by the Kelley brothers (2014).

Project mood: Design agency IDEO seems to be the first one to look at projects in this manner. The exact source or date of this was not found. The illustration is an adaptation of an illustration whose exact source was also not found.

'**Show, don't tell' statement**: Stanford often uses these words for one of its basic principles and various authors refer to this.

'**Ugly does the job done just fine'**: This statement comes from Mike Rohde in a blog about visual thinking, published on January 25, 2011, https://alistapart.com/article/sketching-the-visual-thinking-power-method.

Assignment 'Practicing sketches': This exercise is inspired by a presentation by Supriya Perera at IUX Australia 2014, as described on page 186 of the book Design. Think. Make. Break. Repeat (2018) by Martin Tomitsch and others.

Web example www.advocatenblad.nl: This is part of an article by Nathalie Gloudemans-Voogd for *Advocatenblad* of 24 April 2018, https://www.advocatenblad.nl/2018/04/24/125352/.

Radio advertisement 'You are on the train on your way to work …': The radio advertisement was broadcast on *Radio 1* and the audio clip was found at https://rab.radio/over-rab/rab-commercials/wereldreiziger/.

The structure of a good story: The model is from the book *Imagineering* by Nijs and Peters, page 242.

The elements of a good story: described by Christine Liebrecht in an article on *Marketingfacts*, https://www.marketingfacts.nl/berichten/dit-is-hoe-storytellingwerkt and form a summary of the inaugural speech by José Sanders, professor of narrative communication at Radboud University in Nijmegen.

Guido Stompff Guidelines: These are from the book *Design Thinking*, page 186 through 192.

'Chip in your arm' text box: This is a fictional example based on an anecdotal description of a similar event by proofreader Inge Rijnders.

Quote Arianna Huffington: Arianna Huffington is co-founder and editor-in-chief of the *Huffington Post* and describes the quote in a chapter from the book *The Best Advice I Ever Got* (2011) by Katie Couric. The quote is a life lesson that she learned from her mother, so actually is a quote from her mother Elli Stassinopoulos.

Image 'Good ideas, mediocre ideas, bad ideas': This image is by artist Johan Deckmann, www.deckmann.com.

The myth of the never failing genius: described by the Kelley brothers in their book *Creative Confidence* as 'The Failure Paradox', pages 40 to 42 They refer to an investigation by Dean Keith Simonton; this research is described in his book Origins or Genius: Darwinian Perspectives on Creativity, published by Oxford University Press in 1999.

The benefits of experimenting: The list is compiled on the basis of insights from various authors, including Brown (2009), Brenner, Uebernickel and Abrell (2016) and Stompff (2017).

Drew Houston Quote: This is a ubiquitous quote on the internet; the exact origin was not found.

'Return on learning' text box: This text box is based on a by ABNAMRO sponsored article on nrc.nl: https://www.nrc.nl/advertentie/abnamro/. The quote by Tessa Mulder is also from this article.

Serendipity: The Kelley brothers write about serendipity in *Creative Confidence* on page 105 The Dutch authors Van Andel and Brands have written a whole book about it: *Serendipity*, De Ongezochte Vondst (2014), published by Nieuw Amsterdam Uitgevers.

Text box 'A Nobel Prize by accident?': For the description of the origins of penicillin, Wikipedia was used, https://nl.m.wikipedia.org/wiki/Serendipiteit.

Closing case 'Peka Kroef': The case is based on a project that the author has performed.

2 Design thinking is way of working

Opening case 'Julius Caesar and Emperor Augustus ...': Part of the article 'Julius Caesar and Emperor Augustus could not do without design thinking,' written by Ilse Zeemeijer and published in *Het Financieele Dagblad* on April 29, 2017, https://fd.nl/morgen/1195734/julius-caesar-en-keizer-augustus-konden-niet-zonder-design-thinking.

Design (by) doing: The term 'design doing' is used by many authors, among others, by Tim Brown (2009) and Stickdorn, Hormess, Lawrence and Schneider (2018).

Distinction between the design process and a cycle of design thinking: in literature there is no consensus about this distinction. Flock (2009), Norman (2013) and Herfurth (2016) write about it explicitly. Don Norman writes in *Design of Everyday Things* (page 220) about the Human Centered Design Process as a way to go through the Double Diamond. The author makes the same distinction in the chapter 'Creating and Managing Experiences in Cultural Tourism' from the book *Designing Cultural Tourism Experience* (Jelinčić & Mansfeld) https: // doi.org / 10,1142 / 10809.

Figure 'The design process and the cycle of design thinking': This is an interpretation by the author, based among other things on the design squiggle by Damien Newman, the double diamond by the Design Council and the cycle of design thinking as Guido Stompff describes it. The author's sketch is translated into a concrete figure by graphic designer Tijn Bakker.

Text box 'What does gaming have to do with design thinking?': The example by Fortnite has been checked for truthfulness with the 10-year-old gamer Tieg Ebus.

Assignment 'Play a game': Co-reader Lindsy Szilvasi came up with the tip to do something with Sokoban in the book – based on the Fortnite example.

Figure 'Iterative versus incremental working': Jeff Patton describes it via the Mona Lisa example for the first time in his blog: https://jpattonassociates.com/dont_know_what_i_want/, published January 21, 2008.

Hypotheses according to Bacon: Co-reader Brenda Groen, associate professor at Saxion, referred to the approach to hypotheses by Popper (1959), which is different than how Bacon described it and an article which further elaborates on this: 'A Letter History of the Hypothesis' from Glass and Hal (2008), published in *Cell*, 134 (3), pages 378 through 381.

The cycle of design thinking according to Norman: Don Norman, *Design of Everyday Things* on page 220 about the Human-Centered Design Process.

The cycle of design thinking according to Stompff: Guido Stompff describes the cycle of design thinking in his book *Design Thinking* in chapter 5 (from page 61 on).

Framing and reframing: Guido Stompff defines frames on page 112 in his book *Design Thinking*; he uses the English definition for this by Karl Weick in Sensemaking in organizations (1995), published by Sage.

Text box 'Charli or Jillz?': *Adformatie* dedicated several blogs to the case in 2007: https://www.adformatie.nl/customer-experience/charli-moet-vrouwen-aan-het-bier-krijgen and https://www.adformatie.nl/design/heineken-doopt-charli-om-jillz. That the reason for the name change had to do with the meaning of Charli (cocaine in London street language), seems to come from the British Wikipedia page, https://en.wikipedia.org/wiki/Jillz. Confirmation from Heineken that this is actually one of the reasons for the name change cannot be found.

Ask 'What glasses do you wear?': The information about breakdance has been enriched using Wikipedia, https://en.wikipedia.org/wiki/Breakdance.

Process awareness: Lindsy, working at STUDIO.WHY, Dutch design school, was proofreader of this book and the quote is from one of her comments in Chapter 2.

The double diamond: The double diamond by the *Design Council* developed in 2005. More information about the model can be found at www.designcouncil.org.uk.

Quote Don Norman: The quote is from the book *The Design of Everyday Things* and can be found on page 217.

If **there really is no solution, is there a problem?:** A variation on the last part of a quote from the Dalai Lama: 'If you are confronted with a serious problem, think carefully. If there is a solution, it makes no sense to get all worked up about it. Is there no solution? Then it is no use to get worked up about it.'

Quote 'Love the problem, not your solution': This statement, in all sorts of variants, is used by many design thinkers. Ash Maurya, the author of several books on lean start-ups; for example, uses it consistently. He writes about it in a blog on medium.com, https: //blog.leanstack.com/love-the-problem-not-your-solution-65cfbfb1916b.

He was probably inspired by a statement by Einstein: 'If I were given one hour to save the planet, I would spend 59 minutes defining the problem and one minute resolving it. '

Web example 'www.marketingtribune.nl': This is an edited part of a blog that Robert Buisman wrote for *Marketing Tribune*, published on March 12, 2016, https://www.marketingtribune.nl/b2b/weblog/2016/03/deel-2-uitgelegd-uitgelicht-design-thinking/index.xml.

Action bias: Rolf Dobelli describes the action bias in his book *The Art Of Clear Thinking*. The term action bias and the example of taking penalties are described by Bar-Eli, Azar, Ritov, Keidar-Levin and Schein in the article 'Action bias among elite soccer goalkeepers: *The case of penalty kicks*' (2007), published in *Journal of Economic Psychology*, 28 (5) on pages 606 to 621. They rely, among other things, on research results by Daniel Kahneman and Amos Tversky in 'The Psychology of Preferences '(1982), published in *Scientific American*, 246 (1), page 160 to 173; these researchers describe that 'people feel a more poignant emotional reaction to bad outcomes that result from action relative to otherwise identical outcomes that result from inaction'.

The problem paradox: Kees Dorst writes extensively about the problem paradox in his book *Frame Innovation*. He also uses the example of the fitting room.

Text box 'The paradox of damaged bicycles': The text is based on information from VanMoof.

Design brief: The information about the design brief is partly based on how Google approaches it, https://designsprintkit.withgoogle.com/planning/.

Text box 'Good designers can create normalcy out of chaos': This text is partly based on an article by Jon Kolko from 2010: 'Abductive Thinking and Sensemaking: The Drivers of Design Synthesis, published in *Massachusetts Institute of Technology Design Issues*, 26 (1), page 15 up to and including page 28.

Quote Joost Backus: The author remembers this quote, in German, by Joost Backus: 'Eine Idee ohne form ist nur Potential.' Most likely Joost Backus was inspired by Flaubert: 'Eine Idee ohne Form ist ein Unding, ebenso wie eine Form, die nicht eine Idee ausdrückt.'

Balancing between fluency and flexibility: The Kelley brothers describe this on page 220 of their book *Creative Confidence*.

Idea versus concept: The difference between idea and concept is described in a blog about product design, https://veeel.wordpress.com/2012/03/09/het-verschil-tussen-een-idee-en-een-concept-een-idee-is-waardeloos/; the author is named as 'Riermersma' on this website. The Latin has been checked at https://nl.glosbe.com/la/en/capio.

Text box 'Have all ideas been stolen?': The principle of 'find & modify' comes from Prabir and Chakrabarti in their article 'A Model for the Process of Idea Generation' in *The Design Journal*, 20 (2), pages 239 - 257.

Murder your darlings: Author Forrest Wickman did extensive research for the statement 'Murder your darlings' by Quiller-Couch; the results can be found in his article at http://www.slate.com/blogs/browbeat/2013/10/18/_kill_your_darlings_writing_advice_what_writer_really_said_to_murder_your.html.

Truthfulness prototypes: A blog by Rikke Dam and Teo Siang is used on the website of the Interaction Design Foundation, https://www.interaction-design.org/literature/article/what-kind-of-prototypeshould-you-create, and an article by Menold, Jablokow and Simpson (2017).

Tom and David Kelley quote: a famous saying at IDEO, allegedly by the Kelley brothers.

Boyle's law quote: The author found the quote in a blog by Diego Rodriguez, https://www.linkedin.com/pulse/boyles-law-diego-rodriguez/ on February 3, 2017.

Text box 'Lufthansa: next level prototyping': This case was presented at a conference organized by Köln International School of Design on May 18, 2015 Additional information about the case can be found on the IDEO website, https://www.ideo.com/case-study/elevating-business-class-travel-with-personal connection.

Research Ricardo Martins: Ricardo Martins describes his research on various (service) design methods in *Touchpoint Magazine* 8 (2) page 13.

'**Last-minute adjustments' text box**: This case is based on a project that the author has executed.

Text box 'Arts Center Kubus': This text box is based on a written interview with Myriam Cloosterman, cultural participation consultant at De Kubus.

Text box 'For the reader with fear of flying that still seems sensitive to facts': The investigation into aircraft accidents versus car accidents comes from an RTL News item, https://www.rtlnieuws.nl/economie/column/3793136/factcheck-vliegen-echt-veilig. The calculation for the risk of 'dying during a half-hour car ride is 2.5 times greater than the two hours that you are on the plane to Barcelona' is as follows:

1 billion aircraft kms = 0.04 x 1350 km (Schiphol Amsterdam Airport-Barcelona El Prat) = 54

1 billion cars kms = 0.45 x 30 km = 135; 135/54 = 2.5 times as much chance.

Some tips to promote implementation: from Stickdorn's book, *This is Service Design Doing*, Hormess, Laurence and Schneider (2018) from page 454 on.

Web example 'www.emerce.nl': This is part of a background piece written by Yoshi Tuk on Emerce.nl on May 4, 2017, https://www.emerce.nl/achtergrond/design-thinking-onder-personeel-aansluiten-weer-gedag-zeggen.

'**Samsung' closing case**: This case concerns an adaptation of the item 'How Samsung Became a Design Powerhouse' by Youngjin Yoo and Kyungmook Kim (2015), published in *Harvard Business Review*, 93 (9), pages 73 to 78.

3 Design thinking is a project approach

Opening case 'KLM is investing heavily in design thinking': a post on travelpro.nl from 16 March 2017, written by Arjen Lutgendorff, https: // www.travelpro.nl/video-klm-zet-fors-op-design-thinking/.

Introduction: The use of the roadmaps is partly inspired by Kimbell (2014), who calls this 'recipes' in her book *The Service Innovation Handbook*, published by BIS Publishers.

Quote Robert E. Quinn: This quote concerns the title of the book *Building The Bridge As You Walk On It* by Robert Quinn (2004), published by John Wiley & Sons.

'**Fuzzy front end' text box**: The fuzzy front end in relation to the uncertainty at the start of a project is described by Khurana and Rosenthal in their article 'Towards Holistic Front Ends In New Product Development' (1998), published in the *Journal of Product Innovation Management*, 15, page 57 up to and including 74.

The steps in the discovery phase: The steps in the discovery phase are (partly) based on what Kees Dorst describes in his book *Frame Innovation*: 'steps of frame creation'.

Problem paradox: Kees Dorst writes extensively about the problem paradox in his book *Frame Innovation*.

Classification of stakeholders: This classification is used at Lean/Six Sigma, https://leansixsigmamethodes.nl/wat-is-een-stakeholder.

Interviews: See the 'Interviews' note with the notes in Chapter 4.

5 times why: A blog has been used to describe this, https: //www. toolshero.nl/problem-solving/five-times-why/.

Exercise 'Want to see Rembrandt's The Night Watch up close.': The photo is by photographer Gijsbert van der Wal. Radio maker Botte Jellema went looking for the story behind the photo, read the story behind this controversial photo of the night watch on https://www.nporadio1.nl/cultuur-media/9698-

'**Customer profiles on the street' text box**: The ending up of the customer profiles of Albert Heijn customers was described in *Algemeen Dagblad* by Raymond Boere on January 12, 2018, https://www.ad.nl.

Sunk cost fallacy text box: Kahneman describes various experiments in the book *Thinking, Fast and Slow* on which he bases his findings. The relationship with the Concorde project is described by Arkes and Ayton in their article (1999), published in *Psychological Bulletin*, 125, pages 591 to 600.

Design critique: The information about design critique is partly based on a blog by Scott Berkun, https://scottberkun.com/essays/23-how-to-run-a-design-critique/.

Design brief: The term design brief is widely used; the description is partly based on https://designsprintkit.withgoogle.com/planning/.

Text box 'Don't forget who you do it for': The research mentioned is described in the 'Does a Persona Improve Creativity' article by So and Joo from 2017.

How Might We: How Might We questions are widely used; the description here is partly based on *the Lean Service Creation Handbook* from tech company Futurice.

Quote 'Leigh Thompson': The quote is mentioned in an article by Rebecca Greenfield on the FastCompany website: https: //www.fastcompany.com/3033567/brainstorming-doesnt-work-try-this-technique-instead.

Text box 'How many people should I have the solution tested for?': Don Norman writes about this in his book *The Design of Everyday Things* (2013), on pages 228 and 229.

Closing case 'ITSN': This case is based on a project executed by the author.

4 Design thinking is a tool box

General: The different tools are partly derived from an overview of qualitative from Binkhorst, Den Dekker and Melkert in their contribution 'Blurring boundaries in cultural tourism research' in the publication *Cultural Tourism Research Methods* (pages 41 to 51) by Richards and Munsters (Eds.), Issued by CAB International.

Brainstorming: The information in the introduction comes from https://hbr.org/2015/03/why-group-brainstorming-is-a-waste-of-time. The named steps and variants of brainstorming are partly based on brainstorming techniques in the manual of Marleen van de Westelaken, Vincent Peters and Mario Kieft, available at http://www.samenspraakadvies.nl/publicaties/Handout%20brainstorming%20technieken.pdf.

Business Model Canvas: The Business Model Canvas is mentioned in the book *Business Model Generation* by Alexander Ostenwalder and Yves Pincher. The seven questions come from a blog by Alexander Ostenwalder on his own website http://businessmodelalchemist.com/blog/2011/09/7-questions-to-assess-your-business-model-design.html.

Character profiles: The description of this tool comes from the Design Council website, https://www.designcouncil.org.uk / news-opinion / design-methods-step-3-develop.

COCD box: To describe this tool an article by the editors of Interapreneur.nl was used. The COCD box is conceived by the Center for the Development of Creative Thinking: www.cocd.org.

Customer journey map: Among other things, the information in the book *This is Service Design Doing* by Stickdorn van Hormess, Laurence and Schneider (2018) as from page 44.

Design brief: The information about the design brief is partly based on how Google approaches this: https://designsprintkit.withgoogle.com/planning/.

Design critique: This tool is partly based on a blog by Scott Berkun, http://scottberkun.com/essays/23-how-to-run-a-design-critique/. The addition of '**peer review'** has been suggested by co-reader Brenda Groen.

Desktop walkthrough: The description of this tool is partly based on the article '*Low Threshold Service Design: Desktop Walkthrough*' van Blomkvist, Fjuk and Sayapina (2016).

Focus group: Partly based on Think. Make. Break. Repeat by (among others) Tomitsch, page 64. The addition about the **unfocus group** comes from a blog by Alan Nazarelli, written on August 26, 2016, http://www.siliconvalleyrg.com/svrg-blog/make-your-next-focus-group-an-unfocusgroup.

Highlighter: The highlighter is inspired by the 'Highlights' tool of Booreiland in '*75 tools for creative thinking*' (2012), published by BIS Publishers.

How Might We: How Might We questions are widely used, the description here is partly based on the *Lean Service Creation Handbook* from tech company Futurice.

Interviews: The five steps described are from an Andrews Travers booklet from 2013, *A Pocket Guide to Interviewing for Research*, published by Five Simple Steps.

Observations: The four different ways of observing come from Mike Youngblood's article: 'Four Bedrock techniques for Observational Research 'in *QRCA Views* 12 (2), p. 28-35, 2013.

Personas: The step-by-step plan is based on https://learningpacemethodekit.org/needs-assessment/working-with-data/creating-personas-workshop-method/index.html.

Pitching: The booklet *Pitching Ideas* by Jeroen van Geel was used to describe the tool.

Problem paradox: This tool is based on and is inspired by the description of the problem paradox by Kees Dorst in his book *Frame Innovation*.

Questionnaires: The information used to describe this tool comes from https://www.tvonlinesurveys.com/enquete-maken.

Role play: The description of this tool is partly based on the book *Think. Make. Break.* Repeat from Tomitsch, (among others). The tool is described on page 108 of this book.

Shadowing: Some information has been used to describe this tool by (Spanish-speaking) online community Design Thinking, http://www.designthinking.services/herramientas-design-thinking/shadowing/.

Storyboard: The description of this tool is partly based on the book *Think. Make. Break.* Repeat from Tomitsch (among others) on page 121

User diaries: These tools are based on information from the Design Council at https://www.designcouncil.org.uk/news-opinion/design-methodsstep-1-discover.

Watchtower: The form is based on the 'Immersion' tool from the *Lean Service Creation Handbook* from tech company Futurice.

Workshop: Part of the description of this tool is published in the book Workshops by Nikkie Highmore Sims. The descriptions of the objectives are a free translation of objectives from the Event ROI methodology by Jack Phillips.

About the author

Teun den Dekker (1983) supports organizations in applying design thinking as a way of thinking and working. Since 2013, he has been doing this as creative director and co-owner of /LAB Service & Experience Design. From the start of his career he has been involved as guest lecturer for various courses; in recent years, as *Professor Associat* at Barcelona's Universitat Ramon Llull. He has published articles in international scientific journals on qualitative research methods, co-creation and design thinking.
Contact Teun at: teun@labservicedesign.com.

Index

Illustration credits

Photo's:
Unsplash, Alex Perez: p. 6
Veronica Polinedrio/ Welcome Card: p. 14
Shutterstock: p. 16, 18, 23, 27, 29, 46, 48, 51, 53, 63, 69, 78, 80, 82, 116, 118
White House Press, Washington: p. 25, 26
REX FEATURES/ Shutterstock, New York: p. 32
Unsplash, Javier Allegue Barros: p. 34
Unsplash, Hal Gatewood: p. 37
Maurits Fornier: p. 38
Instagram: @johandeckmann: p. 41
Unsplash, Mike Wilson: p. 42
Danna van Daal-Seuntjens, Peka Kroef: p. 44
Unsplash, Chris Liverani: p. 50
Unsplash, Jannes Glas: p. 58
Unsplash, Jo Szczepanska: p. 59
Unsplash, Ashwini Chaudhary: p. 65
Unsplash, Bernard Hermant: p. 67 (l)
Unsplash, Dan Gold: p. 67 (r)
Gemeente Venlo / Teun den Dekker: p. 72
VanMoof, Amsterdam: p. 77
Gijsbert van der Wal, Zaltbommel: p. 94
Mike Youngblood: p. 175, 176, 177, 177
Claudia Venhorst: p. 240

Technical drawings:
Integra, Pondicherry, India

We have done our utmost to find the copyright holders from whose work we have borrowed extracts or photo's for this specific publication. In case you think you have a copyright infringement issue, we kindly invite you to contact us.

For Product Safety Concerns and Information please contact our EU representative GPSR@taylorandfrancis.com Taylor & Francis Verlag GmbH, Kaufingerstraße 24, 80331 München, Germany

Batch number: 10726426

Printed by Printforce, the Netherlands